Environmental Science and Engineering

Series editors

Ulrich Förstner, Hamburg, Germany
Wim H. Rulkens, Wageningen, The Netherlands
Wim Salomons, Haren, The Netherlands

More information about this series at http://www.springer.com/series/7487

Renata Dulias

The Impact of Mining on the Landscape

A Study of the Upper Silesian Coal Basin in Poland

Springer

Renata Dulias
Faculty of Earth Sciences
University of Silesia
Sosnowiec
Poland

Translated by Elżbieta Madden

ISSN 1863-5520 ISSN 1863-5539 (electronic)
Environmental Science and Engineering
ISBN 978-3-319-29539-8 ISBN 978-3-319-29541-1 (eBook)
DOI 10.1007/978-3-319-29541-1

Library of Congress Control Number: 2016935206

This book is a revised and translated version of: *Denudacja antropogeniczna na obszarach górniczych na przykładzie Górnośląskiego Zagłębia Węglowego*, 2013 published by the University of Silesia in Katowice. The University of Silesia has granted Springer the rights to publish the book.

Printed on acid-free paper

This Springer imprint is published by Springer Nature
The registered company is Springer International Publishing AG Switzerland

Acknowledgments

The author wishes to express her gratitude to Professors Jacek Jania, Leon Kozacki, Maria Łanczont, Zbigniew Podgórski, Andrzej Świeca and Józef Wojtanowicz, who read and corrected the Polish version of this volume, for their constructive comments.

The publication has been partially financed from the funds of the Leading National Research Centre (KNOW) received by the Centre for Polar Studies of the University of Silesia, Poland.

Contents

Chapter 1
Introduction

Mining has accompanied man since the dawn of history. The development of the earliest societies was so closely associated with the use of raw materials that it was reflected in the division of human history into epochs—stone, bronze and iron. Palaeolithic man used over a dozen minerals for making tools and weapons (Coates 1981). More than 30,000 years ago, the walls of the Chauvet cave in the Ardèche valley of France were painted with ochre, and a mine of this residual deposit was recently discovered on the northern coast of Chile. Dating back 12,000 years, it is the oldest evidence of organized mining of any raw material in the Americas (Salazar et al. 2011). Already 6,000 years ago, in underground mines in the Catalan Gavà, a rare mineral called variscite, which was used for ornaments, was extracted (Camprubí et al. 2003). Early cultures of the Middle East in the areas of Israel, Jordan, Turkey, Iraq, Egypt, and Cyprus developed due to agriculture, but it was the dissemination of mining and metallurgy about 6,000 years ago—initially of copper in Anatolia and Israeli Timna in the Arawa valley, and later, tin, silver, and iron—that accelerated the development of these civilizations (e.g. Mannion 2001; Hartman and Mutmansky 2002; Wilkinson 2005). Mining, both local and in subordinate territories, lies at the root of the power of the ancient states of Egypt and Greece and the Roman Empire, mainly Spain, which is extremely rich in raw materials.

Mining generally spread from the Middle East, along the Black Sea and Mediterranean coast, to the west and northwest; thus, in various parts of the Old Continent, it was at different stages of development at the same time. When 7,000 years ago, on the chalk plateau in Belgian Spiennes, Neolithic miners extracted flint in Serbian Rudna Glava and Bulgarian Ajbunar, copper ore was already acquired (Jovanovič 2009; Heeb 2014; Fowler et al. 2015). During the formation of flint mines in Grims Graves on the British Isles and Polish Krzemionki Opatowskie (Allard et al. 2008), the world's oldest salt mine, in Duzdagi in the Azerbaijani Caucasus, already had a thousand years of history (Harding 2013). In ancient Egypt, 6 million tonnes of rock had been hewn to build the pyramids of Cheops (Klemm and Klemm 2001). On the Sudanese island of Meroe, between the

R. Dulias, *The Impact of Mining on the Landscape*,
Environmental Science and Engineering, DOI 10.1007/978-3-319-29541-1_1

Nile and the Atbara, copper ores, gold, and precious stones had been extracted, similar to the Sinai Peninsula—which had been a rich source of copper from Wadi Meggar and Wadi Nash for millennia (Rotenberg et al. 1987).

The sizes of ancient mining centres, which had not been absorbed by desert sands or silted by rivers, indicate that their direct impact on the landscape was generally insignificant on a global scale. Larger areas of preserved post-mining landscape, such as the Spanish Las Medulas—where nearly 2,000 years ago, the Romans mined gold on a large scale using the *montana ruina* method—are relatively few (Jones and Bird 1972). Significant transformations of relief in mining areas began in the Iron Age; they were mainly related to soil erosion on deforested slopes for the purpose of metallurgy, developing on the basis of extracted ores (Mannion 2001; Buchwald 2005). In Middle Ages, besides the mining of various ores, extraction of rock materials had a significant impact on the relief (Hunt and Murray 1999). The biggest impact on the landscape, however, was made by mining of the past 250 years, since the Industrial Revolution in the eighteenth and nineteenth centuries. The scale of modern mining is shown by the following comparison: in the Stone Age, every person consumed no more than a few kilograms of rocks each year for tool making, whereas currently, as many as 9 tonnes are consumed per capita in the United States alone (Nir 1983).

Today, rock materials have fundamental importance in the global mining industry, with their production estimated at about 30 billion tonnes per year, along with coal (6.8 billion tonnes in 2013), oil (3.6 billion tonnes), iron ores (3.1 billion tonnes), and lignite (1.1 billion tonnes). The extraction of 8 raw materials—bauxite, copper ore, manganese, zinc and lead, rock salt, potassium chloride and phosphate rock—in 2013 totalled 0.88 billion tonnes. Almost 60 % of the annual extraction of listed mineral resources come from 5 countries: China, the USA, Russia, Australia, and India; another 13 % come from Brazil, Saudi Arabia, Indonesia, South Africa, and Canada. Despite such a high concentration of mining production—almost three-quarters within 10 countries that cover almost half of the Earth's surface—mining is also important in the economies of most other countries of the world. Mining areas are present on all continents except Antarctica.

Open-pit and underground extraction of raw materials is carried out in different morphoclimatic zones—from oil fields in Alaska in the sub-polar zone, through chromium and nickel mines in the steppes of Kazakhstan, alpine gold mines in the Kyrgyz Tien-Shan, to the Indonesian bauxite mines in the area of rainforests and Moroccan phosphate rock mines in the semi-arid zone. Most of the annual production of basic raw mining materials, however, comes from areas that are located within different varieties of temperate climates (estimated at 40 %), with the least from circumpolar climates (about 3 %) and equatorial climates (approximately 13 %) (Dulias 2013).

Mining activity carried out in such diverse natural conditions raises a variety of environmental problems, which is why numerous research papers on the impact of mining on the environment have been written. The literature on this issue is plentiful—dozens of books have been published in the fields of environmental geology, environmental geomorphology, and industrial ecology (e.g. Coates 1981;

Hester and Harrison 1994; Panizza 1996; Dhar 2000; Ericson 2002; Hummel 2005; Bell and Donnely 2006; Goudie 2006; Spitz and Trudinger 2008) and dozens of scientific articles have been written; a substantial portion of them cover the topic of the impact of mining on the lithosphere and the relief (e.g. Kondolf 1994; Rivas et al. 2006; Sprynskyy et al. 2009).

The extraction of mineral resources is inseparable from the movement of rock masses in the subsurface and the deeper layers of the lithosphere. In areas of intense mining, the amount of displaced material is such that it may not be ignored in the denudation balance. Loss of rock mass caused by mining activity falls within anthropogenic denudation; its increment (increase) is included in anthropogenic aggradation (e.g. Tricart 1960; Zapletal 1969; Demek 1973). A volume comparison of anthropogenic denudation conditioned by mining with natural denudation in most cases reveals at least a several-fold advantage of the former and ranks among the most effective contemporary relief-forming processes (Douglas and Lawson 2001; Wilkinson and McElroy 2007). Research on anthropogenic denudation is therefore one of the most important tasks of modern geomorphology and geology (e.g. Wilkinson 2005; Kirchner and Smolová 2010; Szabó et al. 2010; Price et al. 2011).

In Europe, one of the oldest and largest mining areas is the Upper Silesian Coal Basin located in southern Poland. As a result of long-term coal extraction, but also of stowing sands, zinc and lead ores, and rock materials, the landscape of a significant part of this area has been converted to a degree that allows recognizing it as a model example of an anthropogenic landscape. This work is a comprehensive study of the impact of mining on the landscape of the Basin. The author makes an attempt to answer the question of how mining activity is reflected in the denudation rate of the area. The work includes an outline of the history of mining, the characteristics of anthropogenic relief forms resulting from mining, a description of changes in selected morphometric features of the relief and conditions of the circulation of matter in catchment areas and land-locked basins, the size of anthropogenic denudation calculated on the basis of the raw materials output and on a morphometric analysis, and finally, a comparison of the anthropogenic Upper Silesian Coal Basin landscape with other mining areas in Poland, the Ostrava-Karvina Coal Basin in the Czech Republic, and the Ruhr in Germany.

1.1 Human Activity as a Geomorphic Factor

The problem of landscape transformations due to human activity has been present in the literature for over a hundred years and has undergone successive stages of scientific diagnosis—from ascertaining the fact of transformations and their description, through classifying the forms and processes, to the quantitative and qualitative recognition of transformations. The first work on the impact of human activity on the environment, *Man and Nature*, was written by Marsh (1864). It contained a pioneering insight into the transformation of landscape and created the foundations of anthropogenic geomorphology, although this term was first used by Fels in 1934 in *Der Mensch als Gestalter der Erdoberfläche*. In the meantime, the works of Vojejkov (1894) and Fischer (1915)—*Der Mensch als geologischer Faktor*—were published and, above all, significant works by Sherlock—*Man as a Geological Agent* (1922), *The Influence of Man as an Agent in Geographical Change* (1923) and *Man's Influence on the Earth* (1932), which depicted the problem of transforming the relief by man from the beginning of mankind. Other works by Fels—*Der Mensch als der Erde Gestalter* (1935), and especially *Der wirtschaftende Mensch als Gestalter der Erde* (1954) and *Antropogene geomorfologie* (1957)—are considered to be essential for the development of anthropogenic geomorphology.

Since the mid-twentieth century, most geomorphology textbooks already contained extensive chapters or paragraphs devoted to the anthropogenic relief (e.g. Louis 1960; Klimaszewski 1978; Fairbridge 1968; Demek 1973; Summerfield 1991a; Cooke and Doornkamp 1994; Panizza 1996; Chamley 2003; Migoń 2006; Slaymaker et al. 2009; Gregory 2010). There have also been studies entirely devoted to man as a geomorphological factor (e.g. Zapletal 1969; Nir 1983; Huggett 2003; Szabó et al. 2010; Kirchner and Smolová 2010). Currently, research in the field of *anthropogeomorphology* is conducted in most geographical and geological research centres in the world and relates to different aspects of human pressure on the landscape (e.g. Strahler and Strahler 1973; Rakoczi 1975; Wolman 1975; Haigh 1978; Gregory and Walling 1979; Kadomura 1980; Goudie 1993; Hooke 1999; Häge et al. 1996; Wilkinson 2005; Geist 2006; Rivas et al. 2006; Harnischmacher 2007; Slaymaker et al. 2009; Price et al. 2011; Waters et al. 2014).

The impact of human activity on the relief is both direct and indirect. In the first case, man creates anthropogenic landforms intentionally, such as sand pits, spoil tips, railway embankments, agricultural terraces, or pond dikes. It is man who decides about the location of these forms, their size, shape, time of construction, etc. People also make conscious, concrete steps to change the direction and pace of natural geomorphological processes, such as through the construction of dams retaining debris flow, the stabilization of slopes using concrete ties and metal mesh for protection against landslides, the construction of breakwater to protect cliffs from abrasion of sea waves, and planting trees to prevent aeolian processes, mainly the blowing out of soil.

The indirect human impact on the nature, the course, and the intensity of relief forming processes occur primarily through the impact on vegetation cover. Grubbing and burning of forests, ploughing steppes, and overgrazing farm animals lead to changes to the original land cover and the method of its use. Forests and steppes are converted to crop land, and savannah and steppes into semi-deserts and deserts. It does have an impact on the conditions of the circulation of matter. Although humans do not directly wash away soil from the slopes or dispel it, by destroying the vegetation cover they contribute to this indirectly by creating conditions for increased wind and water impact on land cover. Anthropogenically conditioned revival or weakening of geomorphological processes was initiated a few thousand years ago by the first agricultural civilizations. At present, with the increase of population and the need to meet growing consumption, energy, transportation and other needs, the impact on the landscape has grown to a very large scale. The activity of water or wind accelerated by humans causes soil erosion and exposes bedrock in some areas and leads to the sediment of "anthropogenic" residue, such as colluvia and deluvia or lake sediments, in others. The effect of increased soil erosion comes in the form of rapidly increasing river deltas; for example, the Tigris and Euphrates increased 300 km during the last 4,500 years (Leopold et al. 1964). Indirect effects of anthropogenic pressure on the relief may be revealed at a considerable distance from the point of interference, as well as with a long delay.

Various efforts have been made in an attempt to estimate the global, regional, and local size of anthropogenic pressure and with different time perspectives (e.g. Degens et al. 1976; Nir 1983; Hooke 2000; Wilkinson 2005; Wilkinson and McElroy 2007). The obtained results with the application of different research methods are often not comparable and differ from each other; nevertheless, they always clearly indicate that humans are a very important geomorphological factor, and in some areas their role is even dominant.

Changes in shaping the relief are associated with almost all kinds of human activity—agriculture, mining, industry, transportation, construction, military operations, tourism, and even cults and funeral customs. A number of classifications of anthropogenic landforms, sediments, and processes have been developed. The reader can refer to the ample literature on this subject (e.g. Haigh 1978; Rosenbaum et al. 2003; Ellis et al. 2006; Szabó et al. 2010; Ford et al. 2014). In this study, attention has primarily been focused on the impact of mining on the landscape.

1.1.1 Impact of Mining on the Landscape

Landscape transformations associated with mining were among the first noticeable effects of anthropogenic pressure. Already in 1825, a committee evaluating damage to the relief in the area of Liège stated that it had been caused by the underground extraction of coal (Young and Stoek 1916). In coalfields and ore-bearing areas, which have evolved extensively since the Industrial Revolution, anthropogenic landforms were formed at such a rapid pace that they did not escape the attention of researchers; shortly, humans were considered to be a geological and morphogenic factor (Marsh 1864; Fischer 1915; Sherlock 1922). Currently, the problem of the impact of mining activity on the environment is of interest in many scientific fields and disciplines—mining, geology, hydrology, geography, environmental protection and others.

The intensity of the impact of mining on the landscape is determined by various factors. The first group includes geological factors related to the deposit, such as the type of raw material, the depth of its occurrence, resources, availability, etc. The second group includes factors associated with mining works: the method of extraction, the size of mineral and waste rock extraction, waste production, and the extraction period, etc. The third group includes factors related to landscape features affected by mining: slope inclination, intensity of relief, relative altitudes, etc.

The following types of mining are distinguished: open-pit mining, underground mining, and borehole mining; each of these extraction methods affects the relief directly and indirectly. Direct impacts are based on a conscious, planned occupation of a defined area of land or bedrock for mining activity—the construction of pits, extraction wells and underground workings and the formation of overburden and mining spoil tips. The direct impacts of mining are periodic but often long-lasting. Their unintended side effect is the indirect impact, which includes quakes, and deformation of the surface due to drainage of the deposit, but especially the so-called impact of mining activity carried by the bedrock to surface and causing its deformation.

The greatest direct impact on the landscape is made by opencast mining. According to Nir (1983), most of the world's raw materials (about 70 %) are extracted by opencast mining. This method of extraction is often associated with the implementation of a large-scale, deep mine with several (or even dozens) of mining levels; this is the case of the largest mines in the world, such as in the *Chuquicamata* copper mine in the Atacama Desert and the *Trubka Udacznaja* diamond mine in Yakutia, beyond the Arctic Circle. The pits may take the form of huge cones, vast hollows of regular shape, or irregular workings within the slopes. Numerous but small hollows and mounds constitute remnants of former or illegal mining, such as in the area of Neolithic flint mines at Grimes Graves in Norfolk, England (Ambers 1998). Depending on the type of the extracted raw material, the pits are referred to as open-pit mines, quarries, sandpits, gravel pits, or clay pits. The creation of pits may be associated with the formation of threshold profiles of slopes and even the levelling of entire hills.

In an open-pit mine, exposing a deposit usually requires the removal of overburden, often in large quantities, and storing it on the so-called external spoil tips. The extraction may be accompanied by the production of waste, also stored on spoil tips or in settling tanks. In the case of mining of some raw materials, the quantity of waste is many times greater than the amount of the extracted raw material; for example, the extraction of diamonds in the famous "big hole" in Kimberley brought 3 tonnes of precious stones, but was accompanied by the extraction of 28 million tonnes of soil and rocks (Mannion 2001). Depending on storage techniques and planned development, spoil tips are in the shape of cones, mesas, or piles or they might be irregular. Their height may even exceed 200 m and the occupied area may be as large as several square kilometers. Spoil tips in the depressions may be subsurface or levelling.

In the case of underground mining, pits, shafts, and multi-level systems of galleries are created in the bedrock. Some authors include underground excavations in relief forms as an analogy to the underground karstic forms (Zapletal 1968; Kozacki 1980). The immediate effects of the underground extraction of raw materials are mining spoil tips of waste rock and tailings. These are some of the most characteristic elements of the relief in mining areas (Rainbow 1987;

Harnischmacher 2007), which are morphologically no different from spoil tips associated with open-pit mining.

In any case of deposit extraction, regardless of geological and mining conditions, deformations of relief are formed. They are described in mining of various raw materials: coal (Dunrud and Osterwald 1980; Elifrits et al. 1983; Bullock and Bell 1997; Mc Nally 2000), oil and gas (Poland and Davis 1969; Schoonbeek 1976; Hejmanowski 1993), rock salt (Lee and Sakalas 2001; Cooper 2002; Branston and Styles 2003) and other deposits (Hasan 1996; Galloway et al. 1999; Bell et al. 2002; Becendam 2004; Van Den Eeckhaut et al. 2007), as well as in areas of intense consumption of groundwater (Foose 1967; Poland 1984; Daito et al. 1991; Lebbe 1995; Yong et al. 1995). A variety of mathematical models used to forecast the effects of mining on the bedrock and surface have been developed—for example, based on models of continuous medium, on the theory of stochastic, geometric and integral mediums. However, despite the application of increasingly sophisticated models and computer programmes, taking into account a number of natural, mining and technical parameters, the projected deformations do not always coincide with reality (e.g. Hejmanowski and Malinowska 2009).

The underground extraction of raw materials results in a disturbance of the balance existing in the bedrock. The impact of mining leads to three closely related types of environmental changes—geomechanical, hydrological, and biological. From the geomorphological point of view, the most important are the geomechanical transformations due to the movements of the overlying rocks to the underground workings. They are an inevitable consequence of mining as a prerequisite for the extraction of deposits—the removal of rocks of a certain volume from the rock mass and destruction of the original structure of the bedrock. Geomechanical displacements include increasingly higher rock layers, causing their collapse, cracking, and deflection. Once they get to the surface, they are revealed in the form of continuous deformations (subsidence troughs) and discontinuous deformations (mainly sinkholes, fissures, and thresholds).

The size and distribution of deformations in the bedrock and on the surface depend on a number of factors—both natural (the geological structure above the extracted seam, the depth of the deposit, the thickness, the decline and the number of seams, the hydrogeological conditions, and the occurrence of tectonic dislocations) and mining-engineering (the mining system, the shape, the layout and the progress of the mining front, the number of simultaneously exploited seams, the size of the exploited area, the method of roof management, and the purity of deposit extraction).

The mechanism for creating continuous deformations has been shown in numerous studies (e.g. Bräuner 1973; Borecki 1980; Kratzsch 1983; Holzer 1984; Knothe 1984; Whittaker and Reddish 1989; Peng 1992; Brady and Brown 2004). A major impact on the size and speed of the deformation emergence is made by the overburden thickness and the ratio of the deposit thickness to the overburden thickness—the lower the ratio, the greater are the deformations occurring on the surface. With the depth of extraction, its influence on the surface decreases. Another important factor is also the endurance properties of the rock mass. Hard rocks, such as thick sandstone layers, break up in large sheets, and the resulting surface subsidence trough has gentle slopes and a large range (Osiecki and Trzcionka 1987). Ductile rocks do not crack, even with large deformations, but they strongly deflect; therefore, the subsidence trough has steep slopes and a shorter range. The volume of mining damage is strongly influenced by the dynamics of the extraction—the speed of the mining front, its changes, and the duration of longwall face stoppage. The progress of 2 m/day at a depth of 100 m is comparable with the progress of 20 m/day at a depth of 1,000 m. An important factor influencing the size of the deformation is the extraction system: during a partial extraction (while saving the protective pillars), only insignificant roof subsidence takes place; however, it becomes large during roof caving exploitation. Subsidence coefficients are 0.05 and 0.7, respectively. For stowing exploitation, this coefficient ranges from 0.12 to 0.5 (Borecki 1980).

Continuous deformations are caused by the deflection of overburden rock without breaking the integrity of layers. They come in the form of gentle and extensive subsidence of land called *subsidence troughs*. These are forms of oval or elliptic shape and usually several hundred metres or larger diameter. In the bottom of the troughs, small surface convexities may develop as a result of compression forces liberated during geomechanical transformations. The range of subsidence troughs is greater than the mined-out areas that caused them. The duration of rock mass deformation and relief deformation until reaching the state of equilibrium is several years (3–4); for bedrock built of weak rocks, it is around 1.5 years (Borecki 1980; Whittaker and Reddish 1989).

Geomechanical causes of discontinuous deformations in mining areas have been presented in numerous works (e.g. Beck 1984; Knothe 1984; Whittaker 1985; Karfakis 1987; Singh and Dhar 1997; Wanfang 1997; Wigham 2000; Brown 2003). These deformations are classified mostly due to their shape, distinguishing linear discontinuities (fissures, thresholds) and surface discontinuities (sinkholes). The former are created under the influence of mining activities conducted at different depths and usually accompany continuous deformations. They are frequently formed as a result of the roof caving exploitation of several layers to a border, designated by a protective pillar, a fault, or a mining area. In non-cohesive soils, thresholds are most often formed in cohesive ones—fissures. Linear discontinuities may be erased in the direction towards the relief, especially when they reach loose, sandy layers. A clear impact on abrupt increase rates of deformation and the formation of discontinuities is made by fluctuations of the daily progress of the mining front, mostly by weekend work stoppage (Sroka 2003). The creation of fissures in an area with intense relief may activate landslide processes.

The second type of discontinuous deformations, known as surface deformations, are a result of collapse processes caused by mining activities (Singh and Dhar 1997; Wanfang 1997; Cabala et al. 2004). The threat of surface collapse depends on various factors: the mining factors include the size and the depth of underground workings, modes of operation and roof management, and the type of mine workings casing. Among the geological factors, there is the type and thickness of overburden rocks, tectonic dislocations, and occurrence of the water-bearing rocks. Areas that are particularly vulnerable to collapse are ones with fragile, slit, and porous bedrock (shales, weathered sandstones, limestones, and dolomites).

Most sinkholes (about 80 %) result from shallow and especially roof caving exploitation, and the reactivation of old, shallow workings due to drainage, surface load by structures, or transportation vibrations, reducing the strength of the rocks as a result of rheological processes and weathering, operations on deeper lying seams, or quakes caused by rock burst or landslide movements. The adapted criterion of the likelihood of landslides is the operating depth of up to 80 m, but they are also formed above the more deeply located voids. One of the most important factors activating deformation processes is the circulation of water (natural or anthropogenically forced), such as a result of water pumping in dormant mines. The water that occurs in the overburden initiates the process of tunnelling, causing the displacement of soil particles into the void; therefore, the intensity of the formation of sinkholes increases in times of higher rainfall and decreases in dry or cold seasons (Chudek and Arkuszewski 1980). The emergence of these forms is also contributed to by karst processes.

Surface deformations caused by the collapse of rocks located in the roof of a mine working, such as in the primary void, are generally called sinkholes. An empty space, which is called a secondary void, is formed within the overburden rock; due to subsequent collapses, it "travels" to the surface, reducing volume with regard to the primary void. According to Whittaker (1985), void propagation process to the surface is a *chimney process*. If the secondary void "travels" through a thick rock mass, it may self-stow; however, if it moves to the floor of the loose overburden, it usually results in a sinkhole. Processes prior to its creation may last from several hours to several decades from the formation of a void in the bedrock; however, a void collapse happens rapidly, in a matter of days, hours, or even minutes—therefore, violently. The creation of a deformation is rarely indicated by some characteristic signs. Sinkholes are also formed above filled-in shafts; as a result of the destruction of their casing or a partial washaway of the material by infiltrating water, they may also appear above the voids of old inclined tunnels—declines and galleries.

Sinkholes typically cover small surfaces, causing their destruction; they clearly contrast with the surrounding area, which is hardly subject to any transformations. The continuity of layers between the void and the surface is interrupted. Most discontinuous deformations take the shape of a conical funnel with a depth up to several dozen meters and a diameter exceeding even 100 m (Singh and Dhar 1997; Wanfang 1997) or an irregular sinkhole—shallower, with a flattened bottom. Some smaller forms become silted due to natural geomorphological processes, while some are deliberately liquidated. Sinkhole process modelling is based on the principles of rock mass mechanics and the theory of probability. However, theoretical models are burdened with some flaws. Thus, depending on the adopted theory, the estimated risk of deformations may be disastrous or insignificant (Kotyrba 2005). Forecasting discontinuous deformations therefore relies mainly on the analysis of the occurrence of conditions favourable for their formation.

Indirect effects of underground mining also include seismic events—the so-called rock bursts (Idziak and Zuberek 1995). They cause cracks and sometimes the collapse of buildings; however, in only a few cases, they create fissures in moist soil and landslides of steep slopes and riverbanks. Broader geomorphological consequences arise from mining subsidence. Firstly, due to the lowering of the surface, land-locked basins may be formed, in which the circulation of matter from the previously open changes into closed. If the surface drops below the water-bearing horizon, it results in the flooding and formation of water reservoirs in the depressions. Secondly, surface deformations may cause changes in the hydraulic gradient of rivers and watercourses, and thus lead to the creation of flood plains in valleys and the emergence of a new arrangement of accumulation zones and river erosion (Sidle et al. 2000). Changes in the relief resulting from subsidence may also lead to an imbalance of slopes and consequently initiate landslide processes, especially in areas inherently susceptible to these processes. If underground mining is carried out in a coastal zone, the lowering of the surface (the land and the shelf) influences the course of abrasive processes (Humphries 2001).

Changes accompanying rock mass drainage may also be listed among indirect impacts associated with underground mining (Foose 1967). In some cases, lowering of the surface caused by deposit drainage may exceed the subsidence resulting from its extraction. In addition to compression deformation, there are also deformations of tunnelling or collapse types. An increased formation of sinkholes is observed in the case of deposit drainage in karst areas, as a change in the circulation of groundwater contributes to the tunnelling of sediments covering carbonate rocks (Yu 1994; Wanfang 1997). The forms created in such a manner are called *induced sinkholes* (Newton 1984).

Water pumped out of mines is mostly directed to rivers. Changes in flows—depending on the variable mine water supplies in terms of quantity—modify the processes of erosion and accumulation in river valleys. Mine water, in addition to the loads of salt, carries a suspension that causes the silting-up of sediments and beds. In open-pit mining, indirect influences resulting from draining the excavation are limited by the range of the depression cone (e.g. Dulias 2010). Drainage causes both the lowering of the surface beyond the mine and the disappearance of water (and thus fluvial processes) in the valleys of lesser watercourses.

On the walls of most mines, mass movements occur in the form of screes, rockfalls, or landslides—sometimes on a very large scale. In deep, multi-level excavations, they involve hundreds or even thousands of kilometres of walls (Toy and Hadley 1987). Landslide processes are associated with the disturbance of wall stability, including that caused by the vibration of mining machinery. In mines with water reservoirs, shoreline processes take place (Rzętała M.A. 2003; Rzętała 2008), whereas over-desiccated bottoms of sand pits and other pits where the surface is composed of fine-grained sediment are within the range of aeolian processes (Szczypek and Wach 1991).

A specific type of surface mining is the extraction of raw materials, usually sands and aggregates from the bottom of rivers, water reservoirs, beaches, and coastal areas. In such cases, it is difficult to separate the direct impact from the indirect impact. The exploitation of sediments from river beds causes intensification of deep erosion, changing the geometry of beds, and increased aggradation in the lower course of the river (e.g. Gąsowski 1994; Kondolf 1994, 1997; Mmom and Chukwu-Okeah 2012). Open-pit mining of raw materials in coastal areas enhances abrasive processes (Mensah 1997). As a result of the sand theft procedure in small islands of Indonesia, Malaysia, or Cambodia—sometimes on a massive scale—their surface gradually decreases with the retreat of the coastline. In a similar way, beaches of Mexico, Jamaica, and Morocco are devastated (d'Armagnac 2010). An example of accumulation processes, on the other hand, are deltas built from nickel mine waste in Papua New Guinea (Luick et al. 2011).

Spoil tips, and especially settling tanks, are often subject to aeolian processes (Mizera 1980). At the base of some forms, wavy surface convexities are created due to the pressure of the build-up material on them; this phenomenon is primarily observed on surfaces susceptible to pressure, such as clay (Jankowski 1986). As a result of slopewash processes, building material is carried out to the foreground of spoil tips and other anthropogenic forms; once it joins the river system, in the case of a large supply, it leads to a weakening of river flow (Toy and Hadley 1987).

In summary, the impact of mining on relief consists of the formation of anthropogenic landforms, changes in the morphometric characteristics of the pre-mining relief, modifications of geomorphological processes, or the initiation of new processes. The range of mining impact frequently extends beyond the scope of the operation and the time relationships between the cause and effect are varied—from immediate through delayed to remote, formed after the cease of mining.

1.1.2 Anthropogenic Denudation as a Research Problem

The term *denudation* was introduced to earth sciences in the early nineteenth century. In the first edition of *Principles of Geology* (1832), Lyell used the term *degradation* and eventually described the concept of denudation as "the carrying away by the action of running water of a portion of the solid materials of the land, by which inferior rocks are laid bare" (Lyell 1847 after Gregory 1911). Today, the term *denudation* is not limited to water activity (e.g. Ahnert 1970; Spencer 1983; England and Molnar 1990; Selby 1993; Stüwe and Barr 1998; Ring et al. 1999). According to the definition in the *Encyclopedia of Geomorphology* (2004), "denudation includes all processes that remove the relief at the surface of the Earth." These include physical and chemical weathering, river erosion, glacial and wind erosion, slopewash, and mass movements. Lowering the surface by denudation processes and sedimentation of the denudated material in the concavities lead to the creation of a planation surface.

The quantification of denudation processes is one of the major problems of geomorphology (e.g. Corbel 1959, 1968; Selby 1974; White 1984; Pinet and Souriau 1988; Burbank and Beck 1991; Summerfield 1991a, b; Milliman and Syvitski 1992; Summerfield and Hulton 1994; Goudie 1995; Clayton 1997; Sheen 2000; Walling

2006). A number of different research methods have been developed, but each has its limitations. Denudation indicators are particularly frequently estimated based on the amount of material transported by rivers (rubble, suspensions, dissolved loads), provided that the material is removed from the drainage basins of these rivers (e.g. Degens et al. 1976; Gunnell 1998). The weight or volume of the transported material is divided by the catchment area above the measuring station and expressed in t/km^2/year or mm/km^2/1,000 years. For ease of comparison, Bubnoff units are sometimes applied (B), where 1 B corresponds to 1 mm/1,000 years (Ollier 1981). These indicators inform about the pace of chemical denudation (loss of the rock mass in the form of a dissolved load) and mechanical denudation (loss of the rock mass in the form of a suspension or debris). More reliable results are obtained on the basis of dissolved material, which theoretically comes from the whole catchment area, which cannot be assumed in the case of coarse-grained sediments. Part of the sediments from slope denudation, despite repeated redeposition, remains within it, while some of the material that is incorporated into the river transport may eventually accumulate outside the bed and never leave the catchment. The problems in estimating this "part" are indicated, for example, by extremely different results (75 and 10 %) obtained for the same area—the Amazon basin on the foreground of the Bolivian Andes (Guyot 1993; Mascle and Zubieta-Rossetti 2005). Data interpretation is further impeded by diverse accumulation in a regular year and a flood year.

Denudation indicators are also calculated based on the volume of sediments accumulated in water reservoirs or land-locked basins (e.g. Eardley 1966; Dearing et al. 1982; Foster et al. 1985; Svendsen et al. 1989; Hinderer and Einsele 2001). New capabilities for calculating the rates of denudation are provided by the method based on the use of cosmogenic isotopes (e.g. ^{10}Be, ^{26}Al, ^{36}Cl), producing results broadly in line with other methods (e.g. Cerling and Craig 1994; Harbor 1999; von Blanckenburg et al. 2004).

Since the mid-twentieth century, with the progress of research on the impact of humans on the relief, it has become obvious that humans are an important geomorphological factor; in some areas, the effects of human activities are greater than the effects of other relief-forming processes (Goudie and Viles 1997; Hooke 1999; Price et al. 2011). From a scientific but also practical point of view, it is therefore very important to precisely estimate the contribution of human activity to the total denudation. It has been described as anthropogenic denudation. Tricart (1960), who attributed a very large role to anthropogenic denudation in transforming the relief, referred to the accelerated destruction of arable land, plantations and meadows by depriving them of stable vegetation cover. Denudation is enhanced by agricultural practices, especially when using farm machinery. The consequence of the removal of soil from a slope is the levelling of the surface and accumulation of deposits in the bottoms of river valleys. Zapletal (1968) and Demek (1973) called this process *anthropogenic aggradation*; apart from agricultural causes, they also mention mining activity and levelling or adding soil to the surface for residential, industrial, commercial, transportation and other development. Today, most researchers assume that the movement of rock matter due to different human activities—not only agricultural, but also industrial, construction, water management and, in particular, mining—is included within anthropogenic denudation (e.g. Harnischmacher 2010; Szabó et al. 2010; Dulias 2013; Harnischmacher and Zepp 2014).

It is not simple to estimate the volume of anthropogenic denudation. Unambiguous designation of what part of the total denudation is the denudation caused by anthropopressure most often requires research considering the period prior to human impact on the landscape or parallel research on areas of the transformed and non-transformed environment. Anthropogenic denudation indicators calculated for individual regions or basins are often incomparable due to their environmental diversity (geology, relief, climate, vegetation, etc.). While the anthropogenic denudation scale caused by the direct influence on the relief (e.g., during the exploitation of raw materials) may be reliably calculated, the effects of the indirect impact may only be roughly estimated, without absolute certainty that all circumstances have been taken into account. In the last three decades, the problem of anthropogenic denudation has been discussed in many scientific studies and interesting interpretations but still remains open (Nir 1983; Hooke 1994, 2000; Douglas and Lawson 2001; Rózsa 2007; Wilkinson and McElroy 2007).

1.2 Research Methods and Data Sources

Research on human impact on the landscape of the Upper Silesian Coal Basin has been "positioned" in a short time frame (100–250 years): sensu lato covers the whole area of the Basin, with sensu stricto covering the mining areas of coal mines. The assessment of anthropopressure was carried out according to the same research methods for the whole area. The main elements of research proceedings in succession were the following: literature studies, a compilation of statistical data, gathering of cartographic material, the vectorization of maps, the preparation of digital elevation models, the creation of morphometric databases, and the calculation of indicators of anthropogenic denudation. Field research has also been carried out, including the mapping of selected areas and the observation of contemporary geomorphological processes within selected anthropogenic landforms.

Archival and contemporary cartographic materials were collected on the basis of resources from the Department of Earth Sciences Library of the University of Silesia in Sosnowiec (Poland), the Staatsbibliothek in Berlin (Germany), and the Univerzitní knihovně Ostravské Univerzity in Ostrava (Czech Republic). Having recognized the reciprocal coverage of maps from different centuries, a detailed area of study was designated and the choice of detailed maps for morphometric research was made. The analysis of the Upper Silesian Coal Basin relief from the period prior to mining or from the onset of mining activity was based on Prussian topographic maps of the *Messtischblätter* series on a scale of 1: 25,000 (Messtischblätter 1883), developed on the basis of plain-table survey. These maps, due to their high accuracy and reliability, are considered to be complete cartometric sources of information about the environment and are used in comparative studies of changes of its elements. An average distance error for *Messtischblätter* maps in relation to contemporary topographic maps equals ±41 m—that is, of 1.2 mm on the map scale (Konias 2010). *Messtischblätter* maps provided coverage for the central and western part of the area of detailed research; for the eastern part, for which such maps do not exist, the maps of the Military Geographical Institute *Topographische Karte* on a scale of 1: 25,000 (Topographische Karte 1931) were used, with contour lines of 1881 (Laurahütte sheets, Katowice, part of Imielin) or 1931 (Dąbrowa Górnicza sheets, Birkental, part of Imielin). The analysis of contemporary relief was carried out based on topographic maps on a scale of 1: 10,000 (Mapa topograficzna 1993); in the case of a few areas not covered by these maps, slightly older topographic maps had a scale of 1: 25,000 (Mapa topograficzna 1986).

Cartographic materials (a total of 160 sheets of topographic maps) were rectified in the *ArcMap 9.3* software to the coordinate system of 1992, then vectorized and analysed in the *MapInfoProfessional 7.0*. Digital elevation models were developed for the years 1883 and 1993, each for an area of over 2,800 km^2. By subtracting the contemporary model from the late nineteenth-century model, a raster map of altitude changes during the period 1883–1993 was produced. On its basis, an isoline map was generated, showing areas of decreased and increased heights and areas where heights did not change.

On the basis of digital elevation models, maps of slope inclinations and relative heights were prepared. Databases were created for geomorphological units, for catchments (1st to 6th order), and for mining areas of coal mines. They contain the following data (for 1883 and 1993): the area of different altitude levels above sea level (every 5 m); the maximum, minimum, and relative heights; the inclination of slopes in the intervals every 1°, 2°, 3°, and 5°; the areas of reduced/increased heights in the intervals of 1 or 2.5 m; and the volume of the concave and convex anthropogenic forms, calculated from the subtraction of relief models.

Moreover, the maps inventoried all excavations of minerals and spoil tips legible on the scales of contemporary topographic maps of 1: 10,000 (1993), and their basic dimensions were determined (the size, height or depth, volume). The course of modern watersheds was adopted according to hydrographic maps on a scale of 1: 50,000 (Mapa podziału hydrograficznego Polski 2004). Watersheds of contemporary land-locked basins were determined based on topographic maps of 1: 10,000 (Mapa topograficzna 1993).

Calculations of anthropogenic denudation based on morphometric analysis consisted of dividing the volume of anthropogenic concave forms (direct and indirect) by the research area (the geomorphological unit, catchment, mine) and years of mining activity. The calculated indicator reflects the pace of surface lowering: it was called the anthropogenic denudation rate (D_A) and expressed in mm/year. Anthropogenic aggradation (A_A) was calculated similarly, taking into account the volume of anthropogenic convex forms.

Calculations of anthropogenic denudation based on coal extraction were performed for areas of mines. Mining production volumes were presented using numerous source materials. Various statistical studies of mining and geological institutes, statistical offices, and individual mines were crucial (Statystyka Przemysłu

Węglowego 1945–2009; Rocznik statystyczny Kopalń Węgla Kamiennego 1991–2009; Bilans zasobów kopalin i wód podziemnych w Polsce 1995–2015; Bilans gospodarki surowcami mineralnymi w Polsce na tle gospodarki światowej 1991–2015; Rocznik statystyczny Rzeczypospolitej Polskiej 1998–2009), as well as statistical annexes contained in the works of Luksa (1959), Popiołek (1965), and Jaros (1975). Excel spreadsheets were used to enter data on coal production in all mines of the Upper Silesian Coal Basin (USCB) since the introduction of mining statistics (from 1769) until 2009. Anthropogenic denudation was calculated according to a slightly modified procedure presented by Żmuda (1973). It is a method that allows for an estimate of the size of surface lowering; however, it may not be compared with the results of mining methods that take into account a number of factors affecting the size of surface lowering, including the total thickness of the exploited seams. Waste rock output was estimated by applying the assumption that an average of 0.2 tonnes of waste rock is exploited per 1 tonne of coal (Kupka et al. 2005). Following Żmuda (1973), it was assumed that in the USCB, 1 tonne of coal has a volume of 0.74 m^3 and 1 tonne of waste rock 0.38 m^3; therefore, the volume of the extracted material was calculated (volume of mine workings). Then, based on data regarding the percentage participation of applied operation methods, the volumes of roof caving exploitation and sand-stowing exploitation were calculated. The resulting values were multiplied by the corresponding subsidence coefficients [0.7 for the roof caving and 0.15 for stowing exploitation (Borecki 1980)], then summed and divided by the mining area, followed by the years of the mine's operation. Anthropogenic denudation rates (D_A) were expressed in mm/year. Research periods were established, following the editions of topographic maps used in the morphometric analysis: until 1982, 1983–1993, and 1994–2009.

A significant part of this study is a result of morphometric analysis, with the use of digital elevation models. A comparison of the obtained results with the results quoted in literature and obtained during direct field research revealed some discrepancies between them. Several attempts to verify the results were made and eventually resigned from, preferring the advantage of the research of the entire study area using a single method over the correction of results only possible for certain areas. The indicated differences in results relate mainly to the size of the surface lowering. They are caused by mining, geodetic, and mapping factors, which are presented below.

Among the mining factors, the most important one is the inability to precisely describe all conditions of the bedrock and surface deformation with mathematical formulas. As a consequence, in some areas, subsidence projected by computer programmes differs from the measured and thus actual subsidence (e.g. Popiołek and Ostrowski 1981; Mielimąka 2006; Hejmanowski and Malinowska 2009; Hejmanowski and Kwinta 2010). The projected subsidence error is estimated at 20–30 %, but in rare cases even up to 50 % (Majde et al. 1992). A comparative analysis of subsidence maps, developed for the same area by different research teams, also indicates that they may differ in assessing the scale of the phenomenon by 2–4 m. An example is the area of the Kleofas mine, where the maximum subsidence level is 13 m according to Kupka et al. (2005) or 10 m according to Lasek et al. (2005).

Based on direct field observations, it was further noticed that some of the effects of subsidence were eliminated by surface levelling with artificial soils. This is confirmed by analysis of the anthropogenic ground thickness, based on geological and engineering data (Atlas geologiczno-inżynierski aglomeracji katowickiej 2005). The levelling of subsiding land for residential, industrial, commercial or service development was frequently performed in the area of the Katowice conurbation; therefore, the current height of the area does not reflect the total subsidence of the entire period of mining everywhere.

The results of geodetic measurements are considered the most reliable source of data on surface lowering within a certain period of time; however, there are only a few areas in which long-term and systematic observations were performed (Zych et al. 1994). In the areas of coal mines, geodetic measurements are usually conducted along certain observational lines located in regions that require special protection against mining damages. Surface lowering measured in these sections should not be translated into the whole mining area and especially not undeveloped areas, forests, or agricultural land, where mining activity is not subject to such limitations as in built-up areas; therefore, the subsidence is much greater. In the area influenced by mining activity, geodetic control points are not only reduced but also horizontally shifted (they change coordinates). Maximum horizontal shifts caused by exploitation may reach 40 % of the maximum subsidence in extreme cases; therefore, in conditions of maximum surface subsidence that occurred in the Upper Silesian Coal Basin, in theory, they may amount up to 12 m; in any case, in a substantial part of the USCB, a meter shift may be observed (Majde et al. 1992). High precision demands placed on geodetic measurements therefore necessitate the need for frequent updates of geodetic data in areas that are located within mining exploitation. In recent years, to ascertain the magnitude of subsidence, the applied measurements have used a larger number of benchmarks over a large area (Pomykoł and Kwiecień 1999). Generally, since the publication of Kowalczyk's work (1964), studies of a large-scale deformation in the USCB have rarely been undertaken. Since the 1990s, in the eastern part of the USCB, a uniform, high-precision geodynamic network has been in the process of construction; its aim is to create new opportunities for monitoring displacement points and researching the geodynamics of mining areas (Banasik et al. 2003).

Cartographic materials (archival and contemporary) used in morphometric analysis are considered to be fully cartometric for the study of the geographical environment. Nevertheless, for some areas, the image of relief transformations obtained from the subtraction of digital elevation models for the analysed years (1883, 1993) is burdened with some errors—in areas that have never been located within the exploitation operation zones, both with lowered and increased height (relative to the relief of 1883). Most of these areas have been or still are forested, which raises the presumption of the decreased precision of base maps for less accessible areas. Different accuracy and generalization of the compared maps, resulting from their different scales (1: 25,000 and 1: 10,000) should be remembered. The maximum altitude error for the study area was determined at ±0.45 m, with an average of ±0.16 m. In some areas, as signalled above, the surface was heaped with anthropogenic soils during various construction works; also, subsidence troughs were filled in and the existing spoil tips decreased with an increase in subsidence. This means that not all the calculated values of digital elevation models correspond with the actual volume of changes in the period 1883–1993. A similar opinion was expressed by Wojciechowski (2007).

1.3 Environment of Study Area

The study area is located in southern Poland and includes the Polish part of the Upper Silesian Coal Basin (USCB). Its borders are marked in the west, north, and east by the range of paralic deposits (Doktorowicz-Hrebnicki 1968); in the south by the range of the maximum Carpathian thrust; and in the southwest by the Polish-Czech border. The area of this range covers 4,540 km^2. The detailed study area (sensu stricto) covers the mining areas of coal mines (i.e. the areas assigned to concessions), together with their immediate surroundings. Its borders were determined taking into account the availability and mutual coverage of archival and contemporary cartographic materials. Due to the lack of publication of some sheets of topographic maps, the morphometric analysis did not include 5 mines: the Brzeszcze, Janina, Morcinek, Siersza, and Silesia; three mines were only partially covered: Jaworzno (90 % of the area), Piast (88 %), and Czeczott (33 %). The detailed study area is therefore of irregular shape: it covers 2,838 km^2, of which more than half (1,604 km^2) is the mining areas of the mines. In the work, the adapted names and boundaries of mines, with a few exceptions, are for 1993 due to the need to organize the statistics of extraction—that is, to assign a particular size of coal mining extraction volume to a given area from the beginning of mining operations. The study included mining areas of 63 coal mines, which comprise 90 % of the mines in the USCB. In the last two decades, some mining fields were liquidated and some mines were combined into larger plants; therefore, in 2014, there were 28 active mines open.

The area of the Upper Silesian Coal Basin overlaps with the Upper Silesian Foredeep, belonging to the Variscan structural level (Bukowy 1984). The Foredeep is filled with marine Lower Carboniferous sediments, formed in the Culm or shelly limestone facies and Upper Carboniferous rocks—predominantly sandstones, clays, mudstones and coal deposits. The deposits of Carboniferous productive come in four lithostratigraphic series—the Paralic, the Upper Silesian Sandstone Series, the Mudstone Series and the Cracow Sandstone Series (Jureczka et al. 2005) (Fig. 1.1). The main structural element of the Upper Silesian Foredeep is the Main Syncline. Other outstanding elements are the Main Anticline and the Bytom Syncline in the north, the Wilkoszyn and Chrzanów Synclines in the east and the Jastrzębie Anticline in the west (Buła and Kotas 1994). Carboniferous deposits are predominantly located horizontally and are characterized almost exclusively by disjunctive tectonics—mainly Variscan, but also rejuvenated during the Alpine orogeny. Faults, mostly in the directions of NNE-SSW and WNW-ESE, have discharges

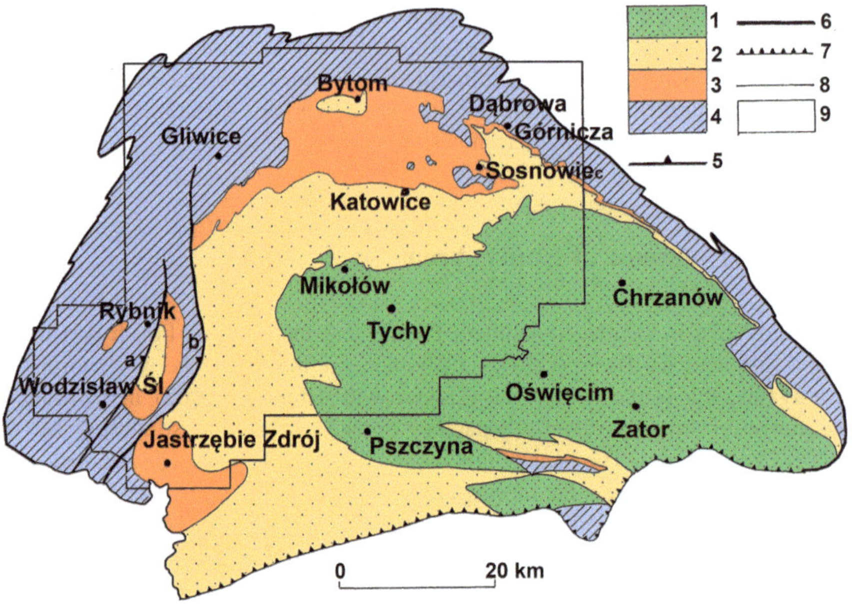

Fig. 1.1 Geological sketch of the Upper Silesian Coal Basin (on the basis of Doktorowicz-Hrebnicki 1968; Jureczka et al. 2005, simplified). (**1**) the Cracow Sandstone Series (Łaziska and Libiąż Beds), (**2**) the Mudstone Series (Załęże and Orzesze Beds), (**3**) the Upper Silesian Sandstone Series (Ruda, Jejkowice and Saddle Beds), (**4**) the Paralic Series (Poręba, Jaklowice, Gruszów and Pietrzkowice Beds), (**5**) thrusts: *a* Michałkowice-Rybnik, *b* Orlova-Boguszowice, (**6**) boundaries of the Upper Silesian Coal Basin after Doktorowicz-Hrebnicki 1968, (**7**) Carpathian thrust, (**8**) the Polish-Czech border, (**9**) detailed study area

from a few to more than 1,000 m; in relief, the particularly visible ones are the Kłodnica fault, the Książęcy fault and the Będzin-Cracow fault. The western part of the Upper Silesian Foredeep margin is characterized by folding tectonics, whereas folding-and-block tectonics are a characteristic of the north-eastern margin.

The overburden of Carboniferous production in northern and eastern parts of the USCB consists of Triassic and Jurassic deposits belonging to the Silesian-Cracow Monocline (Gabzdyl 1994). They are mainly limestones, dolomites and marls; from a mining point of view, ore-bearing dolomites with deposits of zinc and lead ores are of particular importance. The greater part of the Upper Silesian Foredeep (south and west) is covered by Neogene clay sediments belonging to the Carpathian Foredeep. Carboniferous outcrops are located in the central and north-eastern parts of the USCB. The youngest sediments—Pleistocene and Holocene—are genetically (fluvioglacial, glacial, fluvial, aeolian, deluvial) and lithologically (sand, gravel, clay, loess) varied; their thickness reaches several tens of meters. Generally, in the western part of the area, clayey deposits are present; the southern belt features loess; while the centre and in the east have mostly sandy deposits.

The Upper Silesian Coal Basin belongs to two hydrogeological sub-regions. The north-east sub-region includes the Silesian-Cracow Monocline and the Variscan Platform. The aquifers occurring here—Quaternary (porous), Triassic (fissure-karstic) and Carboniferous (fissured-porous)—remain in the hydraulic bond (Różkowski et al. 1997). The south-western hydrogeological sub-region belongs to the Carpathian Foredeep. The Carboniferous aquifer here is isolated by a thick layer of clay and impermeable Tertiary deposits, which is why groundwater only occurs in Quaternary deposits. As a result of intensive drainage of groundwater by mines, a depression cone with an area of about 1,700 km^2 formed, and a natural base of drainage decreased to a depth of 300–700 m. The volume of drained Carboniferous and younger deposits is about 600 km^3; a further 400–500 km^3 is influenced by the drainage (Wilk 2003).

The area of the Upper Silesian Coal Basin is divided into two geomorphological zones. The relief of northern and eastern parts was shaped over a long period of time (about 65 million years) on lithologically diverse Palaeozoic-Mesozoic bed-rock. This area belongs to two macro-regions—the Silesian Upland and the Cracow Upland. Their northern parts are characterized by escarpment relief developed within the Silesian-Cracow Monocline, while the southern parts have horst and graben relief (Fig. 1.2). According to Lewandowski (1993, 1996), the Paleogene planation surface was not prevailed in the contemporary relief, and the main development of escarpment and horst and graben relief took place during the Late Miocene. Structural thresholds and horsts, largely built with chemical denudation susceptible carbonate rocks, are separated by tectonic basins or erosion and denudation depressions. The southern and western part of the USCB was shaped in a relatively short period of time (over a dozen million years) within generally lithologically homogeneous Carpathian Foredeep forms. This area belongs to the Racibórz-Oświęcim Basin. Its morphological diversification is associated with Pleistocene sediments: the western part is a high plain with widely spread clay deposits; in the south-west, there is a loess-covered plateau with numerous ravines; the south is high plains with loess cover; and the remaining part consists of a vast, flat, predominantly sandy depression.

The USCB area has, at least three times, been within reach of the Scandinavian ice sheet—the Sanian I, Sanian II and the Oder. Pleistocene ice and fluvioglacial accumulation concealed the older structural relief, and the degree of its stripping by erosion and denudation processes is spatially varied. At elevations of structural thresholds and horsts, the glaciogenic deposit cover was largely denudated: generally small thickness, colluvial and deluvial deposits occur here; in depressions separating them, there are alluvial and proluvial, fluvioglacial and limnoglacial deposits of several-dozen-meter thickness (Lewandowski 1996).

The contemporary valley network of the USCB coincides with the layout of mining valleys only in its main outlines and refers to the directions of the proglacial outflow formed during the Odranian glaciation. Forced directions of the fluvioglacial draining led to the formation of epigenetic water-gaps during the Eemian interglacial, such as in the valley of the Czarna Przemsza in Będzin. The different height of the erosion base-levels of the Vistula and the Oder is noteworthy (Klimek

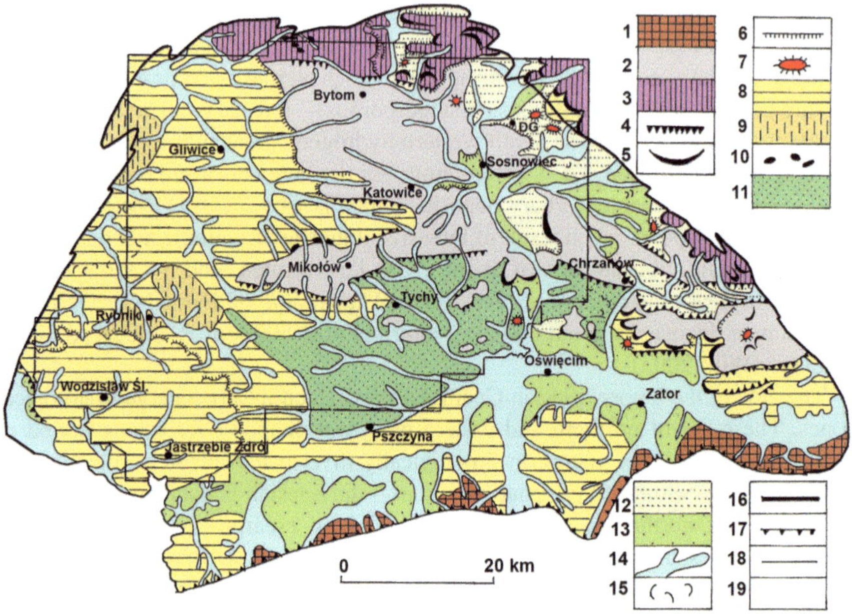

Fig. 1.2 Geomorphological sketch of the Upper Silesian Coal Basin (made by the author on the basis of Przeglądowa Mapa Geologiczna Polski 1980, changed). (**1**) foothills, (**2**) horsts surfaces, (**3**) cuestas surfaces, (**4**) tectonic thresholds, (**5**) cuestas escarpments, (**6**) other erosional-denudational escarpments, (**7**) monadnocks, (**8**) high plains and plateuax with Pleistocene deposits, (**9**) morainic plateaux and high plains with fluvioglacial deposits, (**10**) accumulative moraines, (**11**) outwash plains, (**12**) erosional-denudational depressions with Pleistocene deposits, (**13**) Pleistocene river terraces, (**14**) floors of larger flat-bottomed valleys, (**15**) Late Glacial dunes, (**16**) boundaries of the Upper Silesian Coal Basin after Doktorowicz-Hrebnicki (1968), (**17**) Carpathian thrust, (**18**) the Polish-Czech border, (**19**) detailed-study area, abbreviations: *DG* Dąbrowa Górnicza

and Starkel 1972). Within the study area, the Kłodnica base-level of erosion (the Oder basin) is located 32 m lower than the base-level of erosion of the Przemsza (the Vistula basin). The degree of Holocene rejuvenation of valleys is diversified. In the Racibórz-Oświęcim Basin, valleys located in the Vistula basin are deepened to tens of centimetres, whereas the deepenings of the Oder river basin valleys reaches up to several meters. The bottoms of larger valleys are terraced: the Holocene level is incise by river beds, while the Vistulian terrace progresses into valley slopes, erosion and denudation flattenings, or into sandy outwash plains (Karaś-Brzozowska 1960).

Aeolian activity has been preserved in the landscape through the accumulation of loess during the Vistulian glaciation, mainly in the southern part of the USCB (Dwucet 1986), and the emergence of Late Vistulian and Holocene sand dunes in the valleys of the eastern part (Szczypek and Wach 1993). The relief of the Upper Silesian Coal Basin also bears a clear imprint of anthropogenic transformations, especially on the Bytom-Katowice and Rybnik Plateaus.

The climate in the USCB, because of its large area and varied relief, is quite diversified. An average annual temperature is 7–8 °C. In the distribution of average temperatures in January (from −2 to −4 °C), their decline towards the east is noticeable. Average temperatures in July are, in most parts, 17–18 °C; in the western and eastern margins, they are 18–19 °C (Atlas klimatu województwa śląskiego 2000). Annual precipitation ranges from 630 mm in the Oder valley in the west to 840 mm in the Murcki Plateau in the central part of the Upper Silesian Coal Basin. Dominating winds come from the west; in the Oświęcim Valley, there is a clear dominance of south-west wind. The air is characterized by high dust and gas pollution, especially in the central, highly industrialized, and urbanized part of the USCB.

The larger, eastern part of the USCB (60 % of the area) belongs to the Vistula basin and is drained by the Vistula River, the Czarna Przemsza, the Biała Przemsza, the Pszczynka, the Gostynia, and smaller rivers flowing down from the Cracow Upland, and also the Soła and the Skawa. The western part of the USCB, which is drained mainly by the Kłodnica, the Ruda, the Bierawka, and the Olza, belongs to the Oder basin. The rivers carry water that is heavily contaminated with industrial and municipal waste.

Many different types of soils are represented in the USCB, with a predominance of podzolic and brown soils and the relatively large participation of rendzinas on the weathered carbonate bedrock. A strip of more fertile soils, formed on loess, in the southern part of the USCB is used for agriculture. The highly urbanized areas of the Katowice conurbation in the north of the USCB and the Rybnik conurbation in the south-west are separated by a vast complex of Pszczyna-Ruda forests. There are also numerous forests in the eastern part of the Basin.

The study area was divided into three zones that, in their general outlines, refer to the presence of older bedrock rocks (Fig. 1.3). These zones—conventionally called Carboniferous, Triassic and Miocene—differ in their style of geological structure, the lithology of bedrock, hydrogeological conditions, mineral resources and the consequent degree of mining anthropopressure. The Carboniferous zone covers the central and north-eastern part of the USCB, with deeply drained bedrock as a result of intensive coal mining. Outcrops of Carboniferous forms cover an area of over 516 km^2. In the highly urbanized landscape, sandstone horsts are noticeable, sometimes with capped Triassic rocks separated by tectonic, erosion and denudation depressions. The Triassic zone (slightly over 284 km^2) covers the northern and eastern parts of the USCB, which is an area where the Carboniferous overburden consists of carbonate Triassic, and, to a lesser extent, Jurassic rocks, accumulating large reserves of groundwater and prone to karst processes. This is an area of structural escarpment and horst and graben relief. Mining pressure has covered the Carboniferous bedrock, together with the Triassic overburden and Quaternary deposits. The Miocene zone, which is the largest (almost 860 km^2, or almost 52 % of the studied mining area), covers the southern and western parts of the Basin, where the Carboniferous substrate is found under impermeable Miocene overburden. Underground waters are found in Quaternary sediments and the landscape is

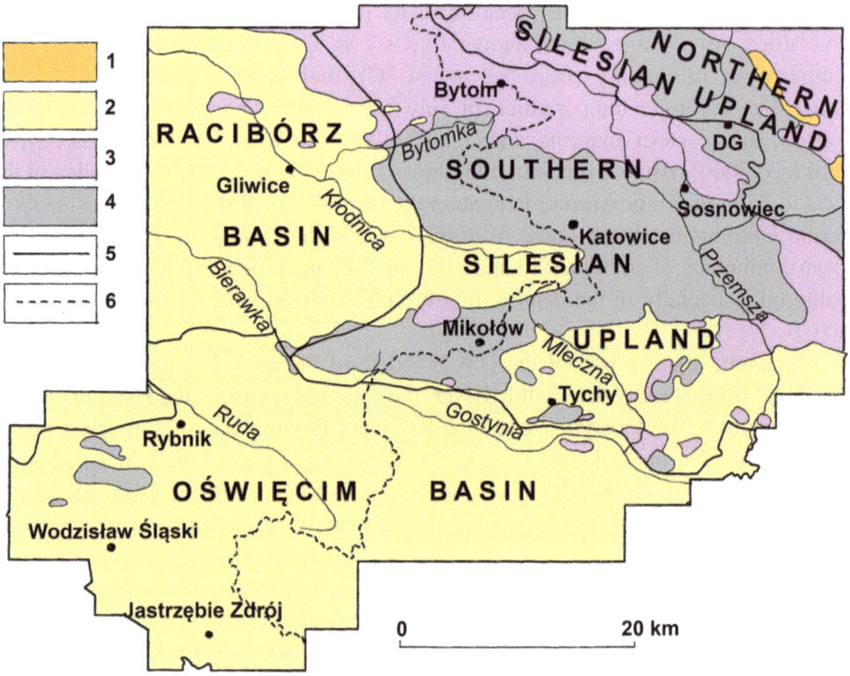

Fig. 1.3 Range of geological zones in the detailed study area. (**1**) Permian deposits, (**2**) Miocene zone, (**3**) Triassic Zone, (**4**) Carboniferous zone, (**5**) boundaries of geomorphological mesoregions, (**6**) the Vistula-Oder watershed; abbreviations: *DG* Dąbrowa Górnicza

dominated by poorly diversified clay high plains, in terms of their relief, and outwash areas with the outstanding Rybnik Plateau, intensely fragmented by gullies and quite heavily urbanized.

References

Ahnert F (1970) A functional relationship between denudation, relief and uplift in large mid-latitude drainage basins. Am J Sci 268:243–263

Allard P, Bostyn F, Giligny F, Lech J (2008) Flint mining in prehistoric Europe. British Archeological Reports, Gordon House, England

Ambers J (1998) Dating Grimes Graves. In: Mok WG, van der Plicht J (eds) Proceedings of the 16th international 14C conference, radiocarbon, vol 40, no 2, pp 591–600

Atlas geologiczno-inżynierski aglomeracji katowickiej 1:10 000. Katowice, Warszawa, Wrocław 2005

Atlas klimatu województwa śląskiego. IMiGW, Katowice, 2000

Banasik P, Góral W, Maciaszek J, Szewczyk J (2003) Wykorzystanie aktywnej sieci geodezyjnej (ASG-PL) do monitorowania przemieszczeń punktów na obszarze GOP. Geodezja 9(2): 169–176

Becendam RF (2004) Stability and subsidence assessment over shallow abandoned room and pillar mines. Lect Notes Earth Sci 104:657–670
Beck BF (ed) (1984) Sinkholes: their geology, engineering, and environmental impact. Balkema, Rotterdam
Bell F, De Bruyn I, Stacey T (2002) Some examples of the impact of metalliferous mining on the environment: a South African perspective. Bull Eng Geol Environ 61(1):1–20
Bell FG, Donnelly LJ (2006) Mining and its impact on the environment. Taylor & Francis e-Library
Bilans gospodarki surowcami mineralnymi w Polsce na tle gospodarki światowej. PAN, Warszawa 1991–2015
Bilans zasobów kopalin i wód podziemnych w Polsce. Ministerstwo Środowiska. PIG Warszawa 1995–2015
Borecki M (ed) (1980) Ochrona powierzchni przed szkodami górniczymi. Wyd. Śląsk, Katowice
Brady BHG, Brown ET (2004) Rock mechanism: for underground mining. Springer, The Netherland
Branston MW, Styles P (2003) The application of time-lapse microgravity for the investigation and monitoring of subsidence at Northwich, Cheshire. Q J Eng Geol Hydrogeol 36(3):231–244
Bräuner G (1973) Subsidence due to underground mining. Part I. Theory and practices in predicting surface deformation. US Department of the Interior, Bureau of Mines
Brown ET (2003) Block caving geomechanics. Univ. J Kruttschnitt Mineral Research Centre, Queensland
Buchwald VF (2005) Iron and steal in ancient times. Historisk-filosofiske Skrifter 29. The Royal Danish Academy of Sciences and Letters
Bukowy S (1984) Struktury waryscyjskie regionu śląsko-krakowskiego. Wyd. Uniw. Śląskiego, Katowice
Buła Z, Kotas A (eds) (1994) Atlas geologiczny Górnośląskiego Zagłębia Węglowego. PIG, Warszawa
Bullock SET, Bell FG (1997) Some problems associated with past mining at a mine in the Witbank coalfield, South Africa. Environ Geol 33(1):61–71
Burbank DW, Beck RA (1991) Rapid, long-term rates of denudation. Geology 19:1169–1172
Cabala J, Teper L, Rutkowski T (2004) Rockmass deformations caused by zinc and lead ores mining in the Olkusz region (southern Poland). Acta Geodyn Geomater 1, 1(133):47–58
Camprubí A, Melgarejo JC, Proenza JA, Costa F, Bosch JA, Estrada A, Borell F, Yushkin NP, Andreichev VL (2003) Mining and geological knowledge during the Neolithic: a geological study on the variscite mines at Gavà, Catalonia. Episodes 26(4):295–301
Cerling TE, Craig H (1994) Geomorphology and in-situ cosmogenic isotopes. Annu Rev Earth Planet Sci 22:273–317
Chamley H (ed) (2003) Geosciences, environment and man. Elsevier, New York
Chudek M, Arkuszewski J (1980) Identyfikacja deformacji zapadliskowych w obszarach dawnej i płytkiej eksploatacji górniczej na terenie Górnośląskiego Okręgu Przemysłowego. Projekty – Problemy Budownictwa Węglowego 4:9–16
Clayton KM (1997) The rate of denudation of some British lowland landscapes. Earth Sur Proc Land 22:721–731
Coates DR (1981) Environmental geology. Willey, Toronto
Cooke RU, Doornkamp JC (1994) Geomorphology in environmental management. Clarendon Press, Oxford
Cooper AH (2002) Halite karst geohazards (natural an man-made) in the United Kingdom. Environ Geol 42:505–512
Corbel J (1959) Erosion en terrain calcaire. Ann Geogr 336(68):97–120
Corbel J (1968) Erozja na powierzchni Ziemi. Studium ilościowe (Metody – Techniki – Wyniki). Przegl. Zagr. Literatury Geogr 23 – Studia nad paleogeografią holocenu, IG PAN, Warszawa, pp 147–180
d'Armagnac B (2010) Le sable marin devient un objet de traffic. Le Monde 29(03):2010

Daito K, Mizuno M, Ueshita K (1991) Control of groundwater withdrawal for preventing land subsidence in the Owari Plain, Japan. IAHS Publ 200:533–542
Dearing JA, Foster DL, Simpson AD (1982) Timescales of denudation: the lake-drainage basin approach. Proceedings of the exeter symposium—recent developments in the explanation and prediction of erosion and sediment yield. IAHS 137:351–360
Degens ET, Paluska A, Eriksson E (1976) Rates of soil erosion. In: Svenson BH, Söderlund R (eds) Nitrogen, phosphorus and sulphur—global cycles. SCOPE Report 7. Ecol Bull (Stockholm) 22:185–191
Demek J (1973) Úvod do studia reliefu Země. SPN, Praha
Dhar BB (2000) Mining and environment. APH Publ. Corp, New Delhi
Doktorowicz-Hrebnicki S (1968) Mapa geologiczna Górnośląskiego Zagłębia Węglowego 1:100 000. Wyd. Inst, Geol, Warszawa
Douglas I, Lawson N (2001) Materials flows for mining and quarrying. In: Munn T (ed) Encyclopedia of global environmental change 3. Causes and consequences of global environmental change, pp 454–461
Dulias R (2010) Landscape planning in areas of sand extraction in the Silesian Upland, Poland. Land Urban Plan 95(3):91–104
Dulias R (2013) Denudacja antropogeniczna na obszarach górniczych na przykładzie Górnośląskiego Zagłębia Węglowego. Wyd Uniw Śląskiego, Katowice
Dunrud C, Osterwald F (1980) Effects of coal mine subsidence in the Sheridan, Wyoming Area. Geoll Surv Prof Paper 1164
Dwucet K (1986) Zróżnicowanie rzeźby na tle litologii utworów pyłowych Płaskowyżu Rybnickiego. UŚ Katowice
Eardley AJ (1966) Denudation rates in plateaus, Southwestern Utah. Geol Soc Amer Bull 77:777–780
Elifrits CD, Barr DJ, Aughenbaugh NB (1983) Room and pillar coal mine subsidence. Geotech Geol Eng 1(4):295–314
Ellis EC, Wang H, Xiao H, Peng K, Liu XP, Ouyang H, Li SC, Cheng X, Yang LZ (2006) Measuring long-term ecological changes in densely populated landscapes using current and historical high resolution imagery. Remote Sens Environ 100(4):457–473
England P, Molnar P (1990) Surface uplift, uplift of rocks, and exhumation of rocks. Geology 18:1173–1177
Ericson J (2002) Environmental geology. Facing the challenges of our changing earth. Facts on File Science Library, USA
Fairbridge R (ed) (1968) The encyclopedia of geomorphology. New York
Fels E (1934) Der Mensch als Gestalter der Erdoberfläche. Petermanns Geographische Mitteilungen
Fels E (1935) Der Mensch als Gestalter der Erde. Bibliographisches Institut AG, Leipzig
Fels E (1954) Der wirtschaftende Mensch als Gestalter der Erde. Franckh`sche Verlagshandlung, Stuttgart
Fels E (1957) Antropogene geomorfologie. Scientia 51:255–260
Fischer E (1915) Der Mensch als geologischer Faktor. Z Deut Geol Ges 67, Berlin, 106–148
Foose RM (1967) Sinkhole formation by ground-water withdrawal, Far East Rand, South Africa. Science 157:1045–1048
Ford JR, Price SJ, Cooper AH, Waters CN (2014) An assessment of lithostratigraphy for anthropogenic deposits. In: Waters CN, Zalasiewicz JA, Williams M, Ellis MA, Snelling AM (eds) (2014) A stratigraphical basis for the Anthropocene. Geol Soc, London, Spec Publ 395:55–89
Foster IDL, Dearing JA, Simpson A, Carter AD, Appleby PG (1985) Lake catchment based studies of erosion and denudation in the merevale catchment, Warwickshire, U.K. Earth Sur Proc Land 10(1):45–68
Fowler Ch, Harding J, Hofmann D (2015) The Oxford handbook of Neolithic Europe. Oxford Univ Press, UK
Gabzdyl W (1994) Geologia złóż węgla. Złoża świata. Polska Agencja Ekol, Warszawa

Galloway DL, Jones DR, Ingebritsen SE (eds) (1999) Land subsidence in the United States. U.S. Geol Surv Circular 1182
Gąsowski Z (1994) L'enfoncement du lit de la Loire. Revue de Géographie de Lyon 69–1:41–45
Geist H (ed) (2006) Our Earth's changing land. An encyclopedia of land-use and land-cover change. Greenwood Press, USA
Goudie A (ed) (2004) Encyclopedia of geomorphology
Goudie AS (1995) The changing Earth. Rates of geomorphological processes. Blackwell, Oxford
Goudie AS (1993) Human influence in geomorphology. Geomorphology 7(1–3):37–59
Goudie AS, Viles H (1997) The Earth transformed. Blackwell, Oxford
Goudie AS (2006) The human impact on the natural environment. 6th edn. Blackwell, Oxford
Gregory JW (1911) The terms denudation, erosion, corrosion, and corrasion. The Geogr J 37 (2):189–195
Gregory KJ (2010) The earth's land surface: landform and processes in geomorphology. SAGE Publ Ltd., London
Gregory KJ, Walling DE (1979) Man and environmental processes: a physical geography perspective. Dawson, Folkestone
Gunnell Y (1998) Present, past and potential denudation rates: is there a link? Tentative evidence from fission-track data, river sediment loads and terrain analysis in the South Indian shield. Geomorphology 25:135–153
Guyot JL (1993) Hydrogéochimie des fleuves de l 'Amazonie Bolivienne. In: ORSTOM (ed) Etudes et thèses 1157–4, 264
Häge K, Drebenstedt C, Angelov E (1996) Landscaping and ecology in the lignite mining area of Maritza-east, Bulgaria. Water Air Soil Poll 91(1–2):135–144
Haigh MJ (1978) Evolution of slopes on artificial landforms, Blainarch, U.K. Univ Chicago, Dep. Geol Res Papers, p 183
Harbor M (ed) (1999) Cosmogenic isotopes in geomorphology. Geomorphology 27:1–172
Harding A (2013) Salt in prehistoric Europe. Sidestone Press, Leiden
Harnischmacher S (2007) Anthropogenic impacts in the Ruhr District (Germany)—a contribution to Anthropogeomorphology in a former mining region. Geogr Fis Dinam Quat 30:185–192
Harnischmacher S (2010) Quantification of mining subsidence in the Ruhr District (Germany). Geomorph Relief Proc Environ 3:261–274
Harnischmacher S, Zepp H (2014) Mining and its impact on the earth surface in the Ruhr District (Germany). Z Geomorph, Suppl Issues 58(3):3–22
Hartman HL, Mutmansky JM (2002) Introductory mining engineering. Wiley, Hoboken
Hasan SE (1996) Subsidence hazard from limestone mining in an urban setting. Environ Eng Geosci 4:497–505
Heeb J (2014) Copper shaft-hole axes and early metallurgy in South-Eastern Europe. Archaeopress Archaeology, Gordon House, England
Hejmanowski R (1993) Zur Vorausberechnung förderbedingter Bodensenkungen über Erdöl-und Erdgaslagerstätten. PhD Thesis Clausthal Technical Univ, Clausthal-Zellerfeld, Germany
Hejmanowski R, Kwinta A (2010) Modelowanie deformacji ciągłych powierzchni terenu w warunkach zmiennego zalegania złoża. Gosp Sur Min 26(3):143–153
Hejmanowski R, Malinowska A (2009) Evaluation of reliability of subsidence prediction based on spatial statistical analysis. Int J Rock Mech Min Sci 46:432–438
Hester RE, Harrison RM (eds) (1994) Mining and its environmental impact. Iss Environ Sci Techn. Royal Society of Chemistry, Great Britain
Hinderer M, Einsele G (2001) The world's large lake basins as denudation-accumulation systems and implications for their lifetimes. J Paleolimnol 26:355–372
Holzer TL (ed) (1984) Man-induced land subsidence. Rev Eng Geol VI
Hooke R LeB (1994) On the efficacy of humans as geomorphic agents. GSA Today 4(9):217–225
Hooke R LeB (1999) Spatial distribution of human geomorphic activity in the United States: comparison with rivers. Earth Sur Proc Land 24:687–689
Hooke R LeB (2000) On the history of humans as geomorphic agents. Geology 28:843–846
Huggett RJ (2003) Fundamentals of geomorphology. Routledge, London

Hummel M (2005) Mining and environment. Technical Univ Ostrava, Ostrava
Humphries L (2001) A review of relative sea level rise caused by mining-induced subsidence in the coastal zone: some implications for increased coastal recession. Climate Res 18:147–156
Hunt ES, Murray J (1999) A history of business in medieval Europe, 1200–1550. Cambirdge medieval textbooks. Cambridge Univ Press, UK
Idziak A, Zuberek WM (1995) Fractal analysis of mining induced seismicity in the Upper Silesia Coal Basin. In: Rossmanith HP (ed) Mechanics of jointed and faulted rocks. Balkema, Rotterdam, pp 679–682
Jankowski AT (1986) Antropogeniczne zmiany stosunków wodnych na obszarze uprzemysłowionym i zurbanizowanym (na przykładzie ROW). Uniw Śląski, Katowice
Jaros J (1975) Zarys dziejów górnictwa węgla. Warszawa
Jones RFJ, Bird DG (1972) Roman gold-mining in north-west Spain, II: working on the Rio Duerna. J Roman Stud 62:59–74
Jovanovič B (2009) Beginning of the metal age in the central Balkans according to the results of the archeometallurgy. J Min Metall 45(2), B:143–148
Jureczka J, Dopita M, Gałka M, Krieger W, Kwarciński J, Martinec P (2005) Atlas geologiczno-złożowy polskiej i czeskiej części Górnośląskiego Zagłębia Węglowego. PIG, Warszawa
Kadomura H (1980) Erosion by human activities in Japan. GeoJournal 4:133–144
Karaś-Brzozowska M (1960) Charakterystyka geomorfologiczna Górnośląskiego Okręgu Przemysłowego. Kom. d/s GOP PAN, Biul 37
Karfakis MG (1987) Chimney subsidence over abandoned coal mines. Geotech Geol Eng 5 (2):131–141
Kirchner K, Smolová I (2010) Základy antropogenní geomorfologie. Univ Palackého, Olomouc
Klemm DD, Klemm R (2001) The building stones of ancient Egypt—a gift of its geology. J African Earth Sci 33:631–642
Klimaszewski M (1978) Geomorfologia. PWN, Warszawa
Klimek K, Starkel L (1972) Kotliny Podkarpackie. In: Klimaszewski M (ed) Geomorfologia Polski 1. PWN Warszawa, pp 117–138
Knothe S (1984) Prognozowanie wpływów eksploatacji górniczej. Wyd Śląsk, Katowice
Kondolf MG (1994) Geomorphic and environmental effects of in-stream gravel mining. Landscape Urban Plann 28:225–243
Kondolf MG (1997) Hungry water: effects of dams and gravel mining on river channels. Environ Manag 21:533–551
Konias A (2010) Kartografia topograficzna państwa i zaboru pruskiego od połowy XVIII wieku do połowy XX wieku. Wyd Nauk Akad Pomorskiej, Słupsk
Kotyrba A (2005) Zagrożenie i ryzyko zapadliskowe terenów Górnośląskiego Zagłębia Węglowego. Wiad Górn 7–8:348–358
Kowalczyk Z (1964) Analiza wyników badań geodezyjnych nad współczesnymi naturalnymi ruchami powierzchni południowej części Górnego Śląska. PAN Geodezja 1, Kraków
Kozacki L (1980) Przeobrażenia środowiska geograficznego spowodowane wgłębnym górnictwem węgla brunatnego na obszarze środkowego Poodrza. UAM, Geografia 21. Poznań
Kratzsch H (1983) Mining subsidence engineering. Springer, Berlin
Kupka R, Szczypek T, Wach J (2005) Morphological effect of 200-years long hard coal exploitation in Katowice. In: Szabó J, Morkūnaitė R (eds) Landscapes—nature and man. Univ Debrecen, Lithuanian Inst. Geol Geogr, Debrecen-Vilnius, pp 95–100
Lasek S, Bubik A, Dygdała W, Lasłowski E (2005) Problem szkód górniczych na terenach likwidowanych kopalń na przykładzie KWK „Katowice-Kleofas". In: Kwiatek J (ed) Problemy eksploatacji górniczej pod terenami zagrożonymi. Ustroń, GIG, pp 345–353
Lebbe L (1995) Land subsidence due to groundwater withdrawal from the semi-confined aquifers of southwestern Flanders. IAHS Publ 234:47–54
Lee EM, Sakalas CF (2001) Subsidence map development in an area of abandoned salt mines. [in:] Land Surface Evaluation for Engineering Practice. Geol Soc, London, Eng Geol Spec Publ 18:193–195

Leopold LB, Wolman MG, Miller JP (1964) Fluvial processes in geomorphology. W.H. Freeman and Company, San Francisco
Lewandowski J (1993) Rzeźba podczwartorzędowa regionu śląsko-krakowskiego i jej ewolucja morfogenetyczna. Folia Quatern 64. Kraków, pp 101–121
Lewandowski J (1996) Główne czynniki neogeńskiej i czwartorzędowej ewolucji morfogenetycznej regionu śląsko-krakowskiego. Acta Geogr Lodz 71:131–148
Louis H (1960) Geomorphologische Wirkungen des Menschen. In: Allgemeine Geomorphologie. Berlin 272–279:315–316
Luick J, Brunskill G, Mudd G, Reichelt-Brushett A, Shearman Ph (2011) Decision nears on Papua New Guinea coastal mine waste dumping. Sci Alert Australia & New Zeland
Luksa J (1959) Rozwój wydobycia w kopalniach węgla kamiennego w Polsce w latach 1769–1948. Stud Mat PTE, Katowice
Lyell C (1847) Principles of Geology. 7th edn. London
Majde A, Nowak E, Śliwka J (1992) Dynamika osnowy geodezyjnej Górnego Śląska. Przegl Geodez 1:11–13
Mannion AM (2001) Zmiany środowiska Ziemi. Historia środowiska przyrodniczego i kulturowego, Wyd Nauk PWN, Warszawa
Mapa Podziału Hydrograficznego Polski 1:50 000. IMiGW Warszawa, 2004
Mapa topograficzna 1:10 000. Główny Geodeta Kraju, 1993–1994
Mapa topograficzna 1:25 000. OPGK, Warszawa 1986
Marsh GP (1864) Man and nature, or physical geography as modified by human action. New York
Mascle G, Zubieta-Rossetti D (2005) Erosion in the Andes and sedimentation in the foreland basin of eastern Boliwia. In: 6th international symposium on Andean Geodyn, ISAG, Barcelona, Extended Abstracts, pp 497–498
Mensah JV (1997) Causes and effects of coastal sand erosion mining in Ghana. Singapore J Tropical Geogr 18(1):69–88
Messtischblätter, Herausgegeben von der Preußischen Landesaufnahme 1:25 000 Reichsamt für Landesaufnahme, Berlin 1883
Mielimąka R (2006) Pomierzone i prognozowane krzywizny terenu górniczego na przykładzie obserwacji geodezyjnych z KWK „Budryk". Górn Geol 1(4):81–92
Migoń P (2006) Geomorfologia. Wyd Nauk PWN, Warszawa
Milliman JD, Syvitski JPM (1992) Geomorphic/tectonic control of sediment discharge to the ocean: the importance of small mountainous rivers. J Geol 100:525–544
Mizera A (1980) Procesy eoliczne na powierzchni zbiornika „Gilów". Cuprum 2:24–27
Mmom PC, Chukwu-Okeah GO (2012) Sand dredging and river morphology change along parts of New Calabar River in Akpor Area of Rivers State, Nigeria and its implications for biological resource conservation. Res J Environ Earth Sci 4(1):82–87
Mc Nally GH (2000) Geology and mining practice in relation to shallow subsidence in the Northern Coalfield, New South Wales. Aust J Earth Sci 47:21–34
Newton JG (1984) Review of induced sinkhole development. In: Beck BF (ed) Sinkholes, their geology, engineering and environmental impact. Balkema Rotterdam
Nir D (1983) Man, a geomorphological agent. An introduction to anthropic geomorphology. D Reider Publ. Co., Boston and Keter Publ House, Jerusalem, Israel
Ollier C (1981) Tektonics and landforms. Geomorphology texts 6. Longman Group Limited, UK
Osiecki A, Trzcionka P (1987) Obserwacje geodezyjne wpływów eksploatacji pokładów zalegających pod grubą ławicą piaskowca. Ochr Teren Górn 80(2):9–12
Panizza M (1996) Environmental Geomorphology. Elsevier
Peng SS (1992) Surface subsidence engineering. Soc Min Metall Exploration
Pinet P, Souriau M (1988) Continental erosion and large-scale releief. Tectonics 7:563–582
Poland JF (ed) (1984) Guidebook to studies of land subsidence due to ground–water withdrawal. UNESCO, PHI Working Group 8.4. Chelsea, Michigan
Poland JF, Davis GH (1969) Land subsidence due to withdrawal of fluids. In: Kiersch GE (ed) Varnes DJ. Rev Eng Geol II, Geol Soc America, pp 187–269

Pomykoł M, Kwiecień D (1999) Próba opisu wpływu eksploatacji górniczej na terenie ROW-u na sieć niwelacji precyzyjnej „Gigant". Zesz Nauk Politech Śląskiej, Górnictwo 239:175–185
Popiołek E, Ostrowski J (1981) Próba ustalenia głównych przyczyn rozbieżności prognozowanych i obserwowanych poeksploatacyjnych wskaźników deformacji. Ochr Teren Górn 58
Popiołek F (1965) Górnośląski przemysł górniczo-hutniczy w drugiej połowie XIX wieku. Śląski Inst Nauk, Katowice
Price SJ, Ford JR, Cooper AH, Neal C (2011) Humans as major geological and geomorphological agents in the Anthropocene: the significance of artificial ground in Great Britain. Phil Trans R Soc A 369:1056–1084
Przeglądowa Mapa Geologiczna Polski 1:200 000. Wyd Geol 1980
Rainbow AKM (ed) (1987) Reclamation. Treatment and utilization of coal waste mining. Elsevier Science Publ, New York
Rakoczi L (1975) Effects of man on sedimentation and erosion in rural environments. Hydrol Sci Bull 20, Washington, pp 103–112
Ring U, Brandon MT, Lister GS, Willet SD (1999) Exhumation processes: normal faulting, ductile flow and erosion. Geol Soc Spec Publ 154:1–27
Rivas V, Cendrero A, Hurtado M, Cabral M, Gimenez J, Forte L, Delrio L, Cantu M, Becker A (2006) Geomorphic consequences of urban development and mining activities; an analysis of study areas in Spain and Argentina. Geomorphology 73(3–4):185–206
Rocznik statystyczny Kopalń Węgla Kamiennego. GIG, Katowice 1991–2009
Rocznik statystyczny Rzeczypospolitej Polskiej. GUS Warszawa 1998–2009
Rosenbaum MS, McMillan AA, Powell JH, Cooper AH, Culshaw MG, Northmore KJ (2003) Classification of artificial (man-made) ground. Eng Geol 69:399–409
Rotenberg B, Bachmann HG, Glass J, Schulman A, Tylecote RF (1987) Pharaonic copper mines in South Sinai. Inst Archaeo-Metall Stud 10(11):1–7
Różkowski A, Chmura A, Siemiński A (eds) (1997) Użytkowe wody podziemne Górnośląskiego Zagłębia Węglowego i jego obrzeżenia. Prace Państ Inst Geol 69, Warszawa
Rózsa P (2007) Attempts at qualitative and quantitative assessment of human impact on the. Geogr Fis Dinam Quat 30:233–238
Rzętała M (2008) Funkcjonowanie zbiorników oraz przebieg procesów limnicznych w warunkach zróżnicowanej antropopresji na przykładzie regionu górnośląskiego. Wyd Uniw Śląskiego, Katowice
Rzętała MA (2003) Procesy brzegowe i osady denne wybranych zbiorników wodnych w warunkach zróżnicowanej antropopresji (na przykładzie Wyżyny Śląskiej i jej obrzeży). Wyd Uniw Śląskiego Katowice
Salazar D, Jackson D, Guendon JL, Salinas H, Morata D, Figueroa V, Manríquez G, Castro V (2011) Early Evidence (ca. 12,000 BP) for Iron Oxide Mining on the Pacific Coast of South America. Curr Anthropol 52(3):463–475
Schoonbeek JB (1976) Land subsidence as a result of natural gas extraction in the Province of Groningen. Soc Petrol Eng AIME, SPE Paper, 5751, Amsterdam
Selby MJ (1974) Rates of denudation. New Zeal J Geogr 56(1):1–13
Selby MJ (1993) Hillslope materials and processes. Oxford Univ Press, Oxford
Sheen SW (2000) A world model of chemical denudation in Karst Terrains. Prof Geogr 52 (3):397–406
Sherlock RL (1922) Man as a geological agent—an account of his action on inanimate nature. Witherby, London
Sherlock RL (1923) The influence of man as an agent in geographical change. Geogr J 61:258–273
Sherlock RL (1932) Man's influnce on the Earth. Home Univ library of modern knowledge, London
Sidle RC, Kamil I, Sharma A, Yamashita S (2000) Stream response to subsidence from underground coal mining in central Utah. Environ Geol 39(3–4):279–281
Singh KB, Dhar BB (1997) Sinkhole subsidence due to mining. Geotech Geol Engin 15:327–341
Slaymaker O, Spencer T, Embleton-Hamann Ch (2009) Geomorphology and global environmental change. Cambridge Univ Press, UK

Spencer EW (1983) Physical Geology. Addison-Wesley Publ Company, Massachusetts
Spitz K, Trudinger J (2008) Mining and the environment: from the ore to metal. CRC Press
Sprynskyy M, Lebedynets M, Sadurski A (2009) Gypsum karst intensification as a consequence of mining activity (Jaziv field, Western Ukraine). Environ Geology 57(1):173–181
Sroka A (2003) Funkcja czasu w świetle wyników obserwacji ciągłych i quasi-ciągłych. Geodezja, AGH 9, 2/1: 507–519
Statystyka Przemysłu Węglowego 1945–2009
Strahler AN, Strahler AH (1973) Environmental geosciences: interaction between natural systems and man. Hamilton Pub Co, Santa Barbara, Calif
Stüwe K, Barr TD (1998) On uplift and exhumation during convergence. Tectonics 17(1):80–88
Summerfield MA (1991a) Global geomorphology: an introduction to the study of landforms. Longman/Wiley, London/New York
Summerfield MA (1991b) Subaerial denudation of passive margins: regional elevation versus local relief models. Earth Planet Sci Lett 102(3–4):4160–4169
Summerfield MA, Hulton NJ (1994) Natural controls of fluvial denudation rates in major world. J Geoph Res 99:13871–13883
Svendsen JI, Mangerud J, Miller GH (1989) Denudation rates in the Arctic estimated from lake sediments on Spitsbergen, Svalbard. Paleogeogr Paleoclimatol Paleoecol 76:153–168
Szabó J, Dávid L, Lóczy D (eds) (2010) Anthropogenic geomorphology. A guide to man-made landforms. Springer, Dordrecht
Szczypek T, Wach J (1991) Human impact and intensity of aeolian processes in the Silesian-Cracow Upland (Southern Poland). Z Geomorph NE, Suppl-Bd 90, Berlin-Stuttgart, pp 171–177
Szczypek T, Wach J (1993) Antropogenicznie wymuszone procesy i formy eoliczne na Wyżynie Śląskiej. SGP, Poznań
Topographische Karte 1:25 000. WIG, Warszawa 1931
Toy JJ, Hadley RF (1987) Geomorphology and reclamation of disturbed lands. Academic Press
Tricart J (1960) Zagadnienia geomorfologiczne. PWN, Warszawa
Van Den Eeckhaut M, Poesen J, Duser M, Martens V, Duchateau Ph (2007) Sinkhole formation above underground limestone quarries: a case study in South Limburg (Belgium). Geomorphology 91:19–37
Vojejkov AI (1894) Vozdeystviya cheloveka na prirodu. Zemlevedeniye 2:4
von Blanckenburg F, Hewawasam T, Kubik PW (2004) Cosmogenic nuclide evidence for low weathering and denudation in the wet, tropical highlands of Sri Lanka. J Geoph Res 109:1–22
Walling DE (2006) Tracing versus monitoring: new challenges and opportunities in erosion and sediment delivery research. In: Owens PN, Collins PN (eds) Soil erosion and sediment redistribution in river catchments. CABI, Wallingford, UK, pp 13–27
Wanfang Z (1997) The formation of sinkholes in karst mining areas in China and some methods of prevention. Environ Geol 31(1/2):50–58
Waters CN, Zalasiewicz JA, Williams M, Ellis MA, Snelling AM (eds) (2014) A stratigraphical basis for the Anthropocene. Geol Soc, London, Spec Publ 395:55–89
White WB (1984) Rate processes: chemical kinetics and karst landform development. Groundwater as a geomorphic agent, Allen and Unwin, pp 227–248
Whittaker BN (1985) Surface subsidence aspects of room and pillar mining. Mining Dept. Magazine, Univ. Of Nottingham 37:59–67
Whittaker BN, Reddish DJ (1989) Subsidence. Occurrence, prediction and control. Dev Geotech Eng 56. Elsevier, Amsterdam
Wigham D (2000) Occurence of mining-induced open fissures and shear walls in the Permian limestones of county Durham. Mining Technology, IMM Transactions section A 109(3): 172–178
Wilk Z (ed) (2003) Hydrogeologia polskich złóż kopalin i problemy wodne górnictwa. Uczel Wyd Nauk Dydakt, Kraków
Wilkinson BH, McElroy BJ (2007) The impact of humans on continental erosion and sedimentation. GSA Bull 119(1/2):140–156

Wilkinson BH (2005) Humans as geologic agents. A deep-time perspective. Geology 33:161–164
Wojciechowski T (2007) Osiadanie powierzchni terenu pod wpływem eksploatacji węgla kamiennego na przykładzie rejonu miasta Knurowa. Przegl Geol 55(7):589–594
Wolman MG (1975) Erosion in the urban environment. Hydrol, Sci Bull 20, Washington, pp 117–126
Yong RN, Turcott E, Maathuis H (1995) Groundwater abstraction-induced land subsidence prediction: Bangkok and Jakarta case studies. IAHS Publ 234:89–97
Young LE, Stoek HH (1916) Subsidence resulting form minning. Univ, Illinois Eng Experiment Station, Bull 91
Yu P (1994) Surface subsidence in the karst mining area in China. Mine Water Environ 13(2):21–26
Zapletal L (1968) Geneticko-morfologická klasifikace antropogennich forem reliéfu. Acta Univ Palackinae Olomunicesis Fac Rerum Natur 23 Geogr Geol VIII:239–427
Zapletal L (1969) Úvod do antropogenní geomorfologie. Univ Palackého, Olomouc
Zych J, Duży S, Kleta H (1994) Wpływ wieloletniej intensywnej eksploatacji górniczej na górotwór i powierzchnię terenu. Mat. Konf - Współczesne problemy ochrony środowiska w górnictwie, Krynica, pp 157–168
Żmuda S (1973) Antropogeniczne przeobrażenia środowiska przyrodniczego konurbacji górnośląskiej. PWN Warszawa-Kraków

Chapter 2
A Brief History of Mining in the Upper Silesian Coal Basin

The area of the Upper Silesian Coal Basin (USCB) is rich in deposits of various mineral raw materials, which, since the Middle Ages have been the subject of mining activities. Galena and silver mining has a thousand-year tradition and it marks the beginning of the economic development of the Silesian-Cracow region. Also, iron ores have been exploited since the Middle Ages. The main importance, in terms of production volume, has been given to coal mining, which has been developing intensely since the Industrial Revolution. At the turn of the nineteenth and twentieth centuries, Poland was one of the leading manufacturers of calamine in the world, thanks to calamine deposits in the USCB. The most important exploitation among the resource-rich rock materials was the exploitation of stowing sands; however, dolomites, limestone, marl, porphyry, melaphyry, and gravel aggregates were also mined. Chemical raw materials found in Miocene deposits, except for short-term mining of gypsum, are not subject to mining activities. Despite plentiful resources, methane is derived from coal beds to a relatively small degree.

In the area of the Upper Silesian Coal Basin, more than 13 billion tonnes of various mineral raw materials have been extracted. The greatest importance has been given to coal mining (80.5 %) and stowing sands (13.3 %). In the total output, the share of aggregate mining (2.3 %) or zinc and lead ores (1.5 %) has been insignificant. The exploitation of other raw materials has accounted for a total of 2.4 % of the total production. The extraction of mineral resources also includes waste rock output in the amount of about 2.1–4.3 billion tonnes.

2.1 Iron Ore Mining

The beginnings of mining in the Upper Silesian Coal Basin date back to the Iron Age and were associated with open-cast exploitation of shallowly deposited bog ores. The fundamental period of ore mining began in the thirteenth century with the

R. Dulias, *The Impact of Mining on the Landscape*,
Environmental Science and Engineering, DOI 10.1007/978-3-319-29541-1_2

exploitation of limonite deposits on outcrops of Triassic rocks, Miocene siderite clay, and Quaternary bog ores. The size of exploitation at that time is difficult to estimate. In the sixteenth century, 8–10 smelt mills operated, which benefited from local ores (Musioł and Płuszczewski 1960); one primitive smelting furnace consumed about 100 tonnes of ores per year (Radwan 1963). It was estimated that in about 300 years (the fifteenth—the seventeenth centuries), about 200,000 tonnes of iron ore were extracted.

In the eighteenth century, ores were exploited for the purposes of 10–12 smelt mills and 5 blast furnaces heated with charcoal (Piernikarczyk 1933/1934). A blast furnace produced less than 100 tonnes of iron per year, and the use of iron from the ore was 60–70 %. For smelt mills, these figures were 32 tonnes and 25–40 %, respectively (Radwan 1963). Limonite ore deposits in the vicinity of Bytom, Tarnowskie Góry, and Piekary Śląskie, accompanying ores of lead and silver in the ore-bearing dolomites, and ores occurring on the surface of Triassic limestones, among karst sediment, were of main importance (Żeglicki 1996). It is estimated that in the eighteenth century, 200–400 thousand tonnes of iron ore were extracted in the Upper Silesian Coal Basin. Traces of this exploitation have been preserved in the landscape to this day, such as in the vicinity of Bytom (Lamparska-Wieland 2003).

In the mid-nineteenth century in the western part of the USCB, in addition to the 20 furnaces heated with charcoal, 8 coke-heated furnaces already operated; this is why the demand for iron ore markedly increased (Musioł and Płuszczewski 1960). Several such furnaces also worked in the eastern part of the USCB in Dąbrowa Górnicza and Sosnowiec. The average annual production during this period amounted to 4,500 tonnes of pig iron for a charcoal-heated blast furnace and 41,000 tonnes for a coke-heated furnace (Czermiński 1992); the use of iron ore was respectively 60 and 80 %. The share of local ores in the blast furnace feed gradually decreased: 70 % in 1878 and only 25 % in 1906. After World War I, the deposits became exhausted and iron ore mining declined in importance. According to estimates for the USCB, a total of about 20 million tonnes of iron ore have been extracted, and the area under the impact of this mining industry covered about 7 km^2.

2.2 Zinc and Lead Ore Mining

Silver and lead ore mining traditions in Poland date back to the beginnings of the Polish state. This was confirmed when a furnace for melting lead was discovered in the north-eastern outskirts of the Upper Silesian Coal Basin, which dates back to the eleventh century (Rozmus and Bodnar 2004). Galena and calamine deposits occur in the Middle Triassic dolomite ore to a depth of about 200 m, occasionally up to 350 m. From the early Middle Ages to the turn of the fourteenth and fifteenth centuries, ores were mined on outcrops in the vicinity of Bytom, Tarnowskie Góry, Jaworzno and Chrzanów. Due to the flooding of mines with groundwater, longer or shorter production breakdowns occurred in ore mining. Exploitation of mineral deposits located below the groundwater table began in the late fifteenth century with the construction of the first dewatering galleries, but still open-pit mining simultaneously continued. The scale of mining at the time is evidenced by thousands of small shafts on the hilltop of the Tarnowskie Góry Hummock, with a total area of approximately 44 km^2. The size of production in this period may only be roughly estimated. By assuming that on average 20,000 centals of lead and silver were produced each year (1 cental = about 58 kg) and considering 150 years of efficient production, then subtracting years of crisis, 174,000 tonnes of metal could be obtained from about 1 million tonnes of ore (assuming about 12 % ore density).

The beginning of the eighteenth century marked the collapse of lead ore mining. However, 100 years later, it was revived with all its intensity in connection with J. Ch. Ruhberg's invention of a method of zinc production from calamine. Then, it was mined in shafts hollowed one next to another, usually to a depth not exceeding 20 m. Large amounts of zinc ore were obtained by sifting and flushing heaps on old mining fields.

The dynamic development of ore mining, which occurred in the mid-nineteenth century, strengthened the division into mining regions: Bytom, Tarnowskie Góry (Bobrowniki) and Chrzanów-Jaworzno (Fig. 2.1). The Olkusz region, sizeable and

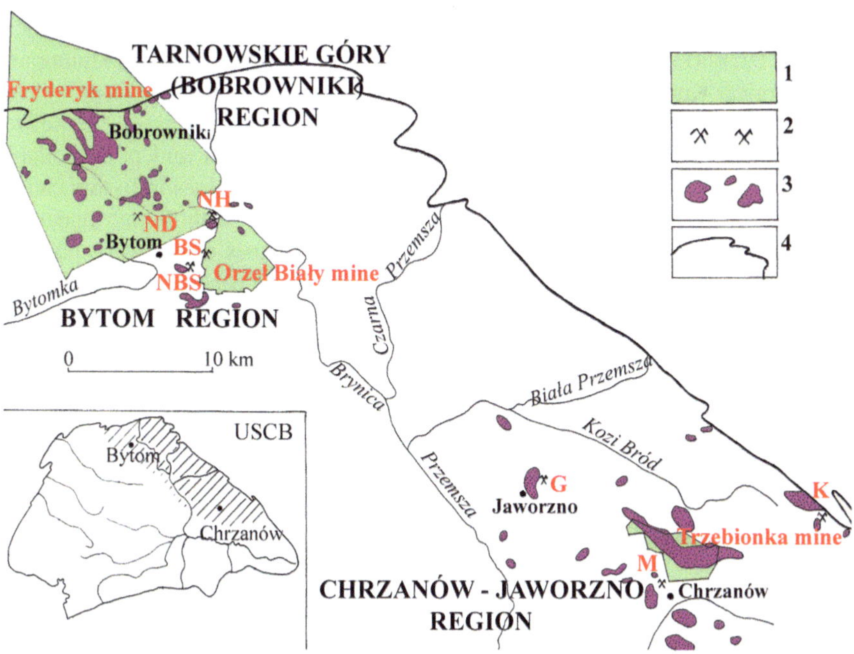

Fig. 2.1 Ore mining in the area of the Upper Silesian Coal Basin (based on Molenda 1972, Messtischblätter 1883). (**1**) Large underground ore mines, (**2**) small ore mines actives in the nineteenth and twentieth centuries: *BS* Biały Szarlej mine, *G* Galmany mine, *K* Katarzyna mine, *M* Matylda mine, *NBS* Nowy Biały Szarlej mine, *ND* Nowy Dwór mine, *NH* Nowa Helena mine, (**3**) main areas of ore mining in the sixteenth–eighteenth centuries, (**4**) north-eastern boundary of the Upper Silesian Coal Basin

important in terms of production, is located outside the USCB. In the Chrzanów-Jaworzno region in 1855, the Matylda mine was established, which operated intermittently due to flooding until 1972. The Katarzyna mine was active until 1912 and the Galmany mine was active until 1958 (Cabała and Sutkowska 2006). The ore mining tradition in this region continued in the years 1962–2009 in the Trzebionka mine, which, at the turn of the twentieth and twenty first centuries, had a reputation as one of the largest in the world, with production of 2.3 million tonnes a year. About 60 million tonnes of ore have been extracted there. In the Chrzanów-Jaworzno region, underground exploitation was carried out in an area of about 15 km^2 and open-pit exploitation in an area of at least 10 km^2.

In the Tarnowskie Góry region, the big lead and silver mine Fryderyk was founded in 1784. It was a state mine; most of it was located within the limits of the Upper Silesian Coal Basin, which was called the Bobrowniki region. In over 126 years of activity (from 1910), this state mine produced 167,000 tonnes of pure lead bullion, whereas the total production of private mines amounted to just over 1 million tonnes (Nowak 1927). Obtaining such a quantity of bullion required the extraction of several million tonnes of ore. Traces of intense mining activity in the Bobrowniki region are still visible in the landscape (Lamparska-Wieland 2003).

In the area of Bytom, 4 zinc and lead ore mines were established: Nowa Helena (1841), Szarlej Biały (1853), Nowy Dwór (1881), and Nowy Biały Szarlej in Bytom (1928) (Fig. 2.1). On the basis of these mines, the Orzeł Biały Mining and Metallurgical Kombinat was founded in 1967, which was one of the largest plants of this type in Europe; its activity ceased in 1989. Zinc and lead deposits in the area of Bytom have been almost entirely exploited with the roof and pillar collapse extraction system (Bąk and Barańczuk 1989). Underground mines in Bytom carried out their operations in an area covering about 18 km^2, with 7 km^2 under the influence of open-pit mining. A total of 22.4 million tonnes of ore were extracted here in the nineteenth century, 32.3 million tonnes in the period of 1901 to 1944, and 48.3 million tonnes in the postwar period until 1984 (Minorczyk 1986).

Since the beginning of ore mining in the Upper Silesian Coal Basin, approximately 200 million tonnes of zinc and lead ores have been output, with 110 million tonnes after World War II. In total, the area under the influence of galena and calamine mining covered about 160 km^2. A significant amount of waste is produced in the process of zinc and lead ore mining and metallurgy—over a hundred million tonnes of it was accumulated. Since 2009, there has been no exploitation of metallic raw materials carried out within the USCB.

2.3 Hard Coal Mining

Hard coal was known to the inhabitants of the contemporary Upper Silesian Coal Basin as early as in the mid-sixteenth century. However, in mining documentation, its exploitation is confirmed only from 1740 at the Murcki mine, which is reported to be the oldest in the Upper Silesian Coal Basin (and still active). The origins of coal statistics date back to 1769. The development of coal mining on the Polish territory occurred in the period of partitions. The borders of the occupying countries divided the USCB into three regions—the Upper Silesia, Dąbrowa and Cracow.

Until the mid-eighteenth century, coal was dug from primitive open pits on outcrops of seams (to the level of groundwater). The total production of pits at that time was only a few hundred tonnes of coal per year. In places of thicker overburden, the deposit was made available through shallow shafts or short galleries; an average depth of operation was only a dozen meters. Mines "wandered" on the ground; the exploitation was conducted within one mining field approximately for 1 year (Kossuth 1961). Coal was mainly used by forges, breweries, distilleries, brickyards, and lime kilns.

The development of coal mining dates back to the founding of the state mines—Król in Chorzów and Królowa Luiza in Zabrze in 1791—as well as the use of coal and coke in ironworks and steam engines that powered drainage pumps in the mines. Primarily, there was a need for bigger lumps of coal; fine coal was not mined for, which was the cause of many fires in the mines. The total coal output by the end of the eighteenth century was relatively small, amounting to 135,690 tonnes; taking into account the numerous small mines (active for a year or up to a few years), it was 308,280 tonnes (Table 2.1).

In the first half of the nineteenth century, in addition to the two large state-owned mines, there were still numerous small underground, sometimes open-pit mines, extracting coal by hand for petty customers. A vast area of shallow coal deposits

Table 2.1 Coal output in large and small mines in the Upper Silesian Coal Basin (USCB) in the years 1769–2009 (based on Luksa 1959; Jaros 1975 and statistical data from the coal industry)

Period	Output in the USCB (tonnes)	Output in large mines (tonnes)	Output in small mines		
			(tonnes)	% of output in USCB	Average (tonnes/year)
To 1800	308,280	135,690	172,590	56.0	5,567
1801–1825	3,796,902	1,832,984	1,963,918	51.7	78,557
1826–1850	15,235,087	6,223,751	9,011,336	59.1	360,453
1851–1875	109,203,807	58,625,284	50,578,523	46.3	2,023,141
1876–1900	445,562,847	364,846,582	80,716,265	18.1	3,228,651
1901–1925	1,013,868,379	951,070,479	62,797,900	6.2	2,511,916
1926–1950	1,497,136,028	1,483,503,421	3,632,607	0.9	545,304
Total 1769–1950	3,085,111,330	2,866,238,191	208,873,139	6.8	1,147,655
1951–1975	2,847,011,739	2,847,011,739	–	–	–
1976–2000	3,962,182,584	3,962,182,584	–	–	–
2001–2009	773,087,900	773,087,900	–	–	–
Total 1951–2009	7,582,282,223	7,582,282,223	–	–	–
Total 1769–2009	10,667,393,553	10,667,393,553	–	–	–

allowed for the transfer of operation along the outcrops of seams over a long period of time and not reaching deeper than the level of dewatering galleries, whose number was at least 75. The number of mines was variable but generally increasing: in the district of Upper Silesia in 1816, there were 27; in 1850, there were 71 (Luksa 1959). In order to increase profits, only thick layers were extracted, with thickness greater than 1 m. The main consumer of coal was zinc metallurgy. In the first half of the nineteenth century, 19 million tonnes of coal were extracted in the Upper Silesian Coal Basin, of which as many as 11 million tonnes were in small mines (Table 2.1).

In the second half of the nineteenth century, the development of coal mining was encouraged by the construction of railway lines, the concentration of capital and mine mergers, and rapid technological progress. Rail transport not only enabled the export of coal, but it also became its important recipient. Mines were located in a strip stretching from Zabrze in the west to Mysłowice in the south-east, along Saddle Beds, shallowly located and characterized by exceptional thickness and small or medium inclination. A smaller concentration of mines was formed in the vicinity of Rybnik in the south-western part of the USCB. During this period, mine drainage became a necessity. Stowing (slag, ash, sand, heap waste) began to be used occasionally in the late nineteenth century. Over 40 % of the output came from the mining of seams thicker than 4 m, mainly by the so-called Silesian pillar system. In the eastern part of the Basin, the Dąbrowa method was applied, which was the extraction of the entire thickness of the seam. Since 1857, in the Król mine and several others, thick layers were extracted by the checkerboard method, which resulted in very large losses of coal, even up to 40–50 %. Overall, the management of thick coal in Upper Silesia was wasteful. The dominating method was roof collapse excavation, which led to the destruction of large coal deposits in thinner seams (Kossuth 1965). The depletion of shallow deposits led to the stabilization of mines in the area: their number decreased by half in the late nineteenth century—from a maximum of 142 in 1873 (Popiołek 1965). In total, in the second half of the nineteenth century, 554.8 million tonnes of hard coal were extracted.

In the early twentieth century, coal production had already reached a high level of concentration. The depth of operation generally did not exceed 100 m (Jaros 1969). Still, mostly thick seams were exploited; seams thinner than 1.8 m were only exploited in exceptional cases (Kossuth 1968). Fine coal, which had formerly been wasted on heaps, started to be divided on assortments and purified in flushers. Hydraulic stowage was used for the first time in 1901 on a larger scale in the Mysłowice mine and in the following years in many other mines in Upper Silesia. On the eve of World War I, there were 103 active mines. Large mines in the Upper Silesia region (63) at the time produced more than 80 % of total output for the entire Basin. In the Dąbrowa region, half of the 32 mines were classified as small and shallow, while 8 coal mines in the Cracow region accounted for only a few percent of the whole mining production in the whole USCB (Jaros 1969). The largest war losses were experienced by the mines in the Dąbrowa region. The curve of coal production in the inter-war period reveals a series of leaps and crashes—periods of economic recovery were separated by periods of crisis, when some mines were closed down. Unemployed miners illegally mined coal in illegal (poverty) pits, particularly in the areas of Dąbrowa Górnicza, Katowice, and Świętochłowice; in 1932, there were about 5,000–6,000 of these illegal pits in the region (Ziemba 1967).

During World War II, mines became included in the framework of German war economy and coal production greatly increased. During the 1944–1945 offensive, Polish mines suffered no serious damage, but the technical and mining condition of mines after the war was poor—the casing of workings was 90 % wood, and loading of the extracted material was carried out almost exclusively by hand (Górnictwo 1988). Almost half of the extraction came from a depth up to 300 m, and 42 % came from seams thicker than 3.5 m. About 70 % of extraction was roof-caving exploitation (Jaros 1973). In the first half of the twentieth century, 2.5 billion tonnes of hard coal were extracted (Table 2.1).

After World War II, for many years, coal constituted the foundation of the Polish energy sector and steel industry and played an important role in the balance of trade. Initially, the exploitation was concentrated in the central part of then the Upper Silesian Industrial Region (USIR). In 1947, a total of 66 mines were active, most of which were small, with the extraction not exceeding 1 million tonnes (Tkocz 1987). Due to the high demand for coal, several new mines were built in the USIR (e.g. Halemba, Staszic); in the 1960s and 1970s, many mines were founded in the Rybnik Coal District (e.g. Jastrzębie, Pniówek) in bedrock with particularly difficult geological and mining conditions—highly disturbed seams and one of the highest coalbed methane hazards in Europe (Górnictwo 1988). In total, in the postwar period, 22 new underground mines and 33 open-pit mines were established, including the largest—Brzozowica. The newest mine is Budryk. The locations of coal mines in the Upper Silesian Coal Basin are shown in Fig. 2.2.

In 1974, most mines had an average extraction of 2–4 million tonnes; 10 years later, 13 mines extracted more than 5 million tonnes, including Ziemowit and Piast with more than 7 million tonnes (Tkocz 1987). In the late 1980s, the USCB supplied 98.5 % of the Polish coal production. The increase in coal output was related to the technological revolution in Polish mines initiated by the introduction of mining shearers for wall exploitation: in 1950, there were 6 of them, whereas in 1985, there were 765. The main exploitation system became the exploitation of longwall faces, with the roof-collapse method or hydraulic stowage. After the war, the problem of excavating seams under whole cities, primarily under Bytom, was academically researched (Krupiński 1956). In 1952, experimental exploitation of the protection pillar under the Pokój steelworks was initiated (Skinderowicz 1963); later, pillars under the city of Bytom, steel mills, and some shafts were exploited. In the 1960s, about 35 % of extraction came from protective pillars; since 1966, some mines achieved most of their production from them, such as the Pokój mine (Jaros 1973). The recovery of millions of tonnes of coal, which had formerly been thought to be lost, had its price in the form of vast mining damages on the surface. Coal output at individual mines for the years 1769–2009 is shown in Table 2.2.

As a result of the restructuring of mining, since the 1990s, unprofitable mines were closed down or merged into larger mining plants. In 2014, there were 27

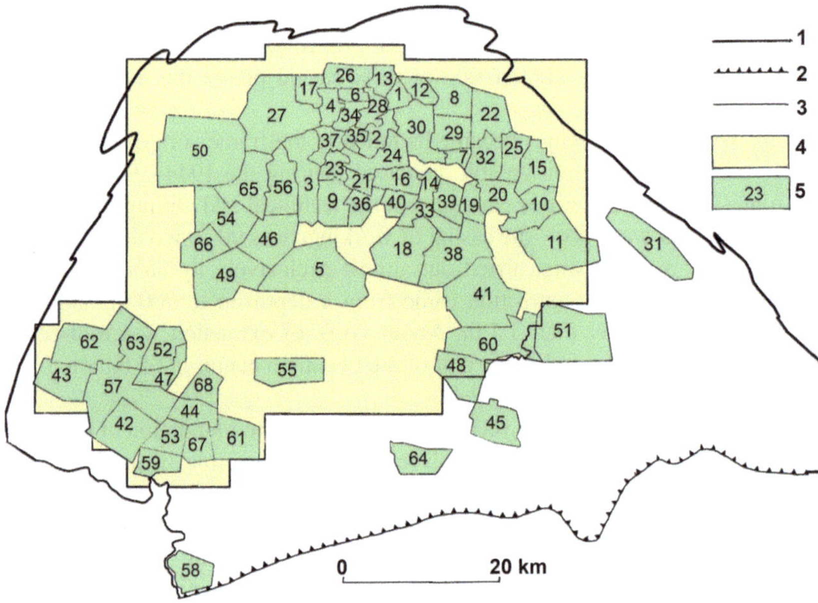

Fig. 2.2 Mining areas of coal mines in the Upper Silesian Coal Basin in 1993 (after Dulias 2013). (**1**) Boundaries of the Upper Silesian Coal Basin after Doktorowicz-Hrebnicki 1968, (**2**) Carpathian thrust, (**3**) the Polish-Czech border, (**4**) detailed-study area, (**5**) coal mines (the numbering of mines consistent with the numbering in Table 2.2): *1* Andaluzja, *2* Barbara-Chorzów, *3* Bielszowice, *4* Bobrek, *5* Bolesław Śmiały, *6* Centrum, *7* Czeladź-Milowice, *8* Grodziec, *9* Halemba: *10* Jan Kanty, *11* Jaworzno, *12* Jowisz, *13* Julian, *14* Katowice, *15* Kazimierz-Juliusz, *16* Kleofas, *17* Miechowice, *18* Murcki, *19* Mysłowice, *20* Niwka-Modrzejów, *21* Nowy Wirek, *22* Paryż, *23* Pokój, *24* Polska, *25* Porąbka-Klimontów, *26* Powstańców Śląskich, *27* Pstrowski, *28* Rozbark, *29* Saturn, *30* Siemianowice, *31* Siersza, *32* Sosnowiec, *33* Staszic, *34* Szombierki, *35* Śląsk-Matylda, *36* Śląsk, *37* Wawel, *38* Wesoła, *39* Wieczorek, *40* Wujek, *41* Ziemowit, *42* 1 Maja, *43* Anna, *44* Borynia, *45* Brzeszcze, *46* Budryk, *47* Chwałowice, *48* Czeczott, *49* Dębieńsko, *50* Gliwice, *51* Janina, *52* Jankowice, *53* Jastrzębie, *54* Knurów, *55* Krupiński, *56* Makoszowy, *57* Marcel, *58* Morcinek, *59* Moszczenica, *60* Piast, *61* Pniówek, *62* Rydułtowy, *63* Rymer, *64* Silesia, *65* Sośnica, *66* Szczygłowice, *67* Zofiówka, *68* Żory

active mines. Currently, the longwall exploitation method is used, predominantly by roof-collapsing (85 %). The lengths of the walls are 100–200 m (up to 275 m), and their progress reaches an average of 40 m per month (up to 200 m in the Staszic mine; Kowalski 1996). The number of active mining walls, however, decreased from 766 in 1999 to 142 in 2004, but an average output from the wall increased from 863 tonnes a day (1990) to 2,920 tonnes a day (2004) (Karbownik and Włodarski 2005). Some shafts of the liquidated coal mines have been transformed into deep well shafts, which pump groundwater out to remove the water hazard in active mines. Cessation of mine dewatering and restoration of the natural groundwater level would create flooding and water reservoirs in areas of major subsidence, including the centres of Silesian cities (Kotyrba 2005).

Table 2.2 Coal output in the mines of the Upper Silesian Coal Basin (USCB) in the years 1769–2009 (based on Luksa 1959 and statistical data from the coal industry)

No.	Coal mines (named by the state in 1993)	Coal output								Year of start of mining	Years of activity since the begining of miming statistic
		to 1882[a]		1883–1993		1994–2009		Total	% of output in USCB		
		mln tonnes	%	mln tonnes	%	mln tonnes	%	mln tonnes			
Silesian Upland											
1.	Andaluzja	–	–	131.5	84.2	24.7	15.8	156.3	1.49	1911	99
2.	Barbara-Chorzów	5.9	4.3	132.2	95.7	–	–	138.1	1.34	1791	125
3.	Bielszowice	19.0	4.1	396.4	85.4	48.8	10.5	464.3	4.44	1791	213
4.	Bobrek	–	–	165.1	86.1	26.7	13.9	191.8	1.83	1907	100
5.	Bolesław Śmiały	2.2	1.2	144.2	82.9	27.6	15.9	173.9	1.66	1779	205
6.	Centrum	–	–	151.1	85.7	25.2	14.3	176.3	1.69	1865	109
7.	Czeladź-Milowice	1.1	0.8	140.3	98.7	0.7	0.5	142.1	1.36	1822	174
8.	Grodziec	–	–	52.0	93.1	3.9	6.9	55.9	0.53	1823	105
9.	Halemba	–	–	116.7	69.1	52.3	30.9	169.0	1.62	1954	56
10.	Jan Kanty	–	–	70.5	92.2	6.0	7.8	76.5	0.73	1766	81
11.	Jaworzno	5.1	2.0	216.5	83.0	39.3	15.0	261.0	2.50	1795	214
12.	Jowisz	–	–	85.9	94.3	5.2	5.7	91.1	0.87	1912	89
13.	Julian	–	–	79.5	79.2	20.9	20.8	100.4	0.96	1955	55
14.	Katowice	1.7	1.4	112.9	92.7	7.2	5.9	121.8	1.16	1822	133
15.	Kazimierz-Juliusz	–	–	115.4	88.4	15.1	11.6	130.5	1.25	1875	116
16.	Kleofas	–	–	178.5	91.5	16.6	8.5	195.2	1.87	1845	117
17.	Miechowice	–	–	128.7	98.4	2.2	1.6	130.9	1.25	1902	94
18.	Murcki	2.4	1.5	119.6	74.3	39.0	24.2	161.1	1.54	1740	240

(continued)

Table 2.2 (continued)

No.	Coal mines (named by the state in 1993)	Coal output								Year of start of mining	Years of activity since the begining of miming statistic
		to 1882[a]		1883–1993		1994–2009		Total	% of output in USCB		
		mln tonnes	%	mln tonnes	%	mln tonnes	%	mln tonnes			
19.	Mysłowice	0.2	0.1	135.5	82.7	27.8	17.2	161.5	1.54	1837	137
20.	Niwka-Modrzejów			101.4	91.8	9.0	8.2	110.4	1.06	1815	106
21.	Nowy Wirek	3.9	3.5	105.7	95.3	1.4	1.2	110.9	1.06	1849	123
22.	Paryż	6.6	4.2	150.3	95.4	0.6	0.4	157.5	1.51	1785	211
23.	Pokój	1.0	0.5	157.6	83.1	31.0	16.4	189.6	1.81	1865	145
24.	Polska	23.3	7.8	251.2	83.8	25.0	8.4	299.6	2.86	1791	219
25.	Porąbka-Klimontów	–	–	148.0	97.5	3.8	2.5	151.8	1.45	1806	105
26.	Powstańców Śl.	0.5	0.2	223.7	95.7	9.5	4.1	233.8	2.24	1871	125
27.	Pstrowski[b]	6.0	1.5	390.5	97.5	4.0	1.0	400.5	3.83	1792	148
28.	Rozbark	5.9	2.8	192.6	89.7	16.0	7.5	214.6	2.05	1824	143
29.	Saturn	–	–	74.8	99.4	0.5	0.6	75.3	0.72	1887	109
30.	Siemianowice	6.3	2.0	298.4	96.1	5.8	1.9	310.5	2.97	1788	212
31.	Siersza	1.5	1.0	139.9	94.8	6.2	4.2	147.7	1.41	1804	196
32.	Sosnowiec	1.9	1.6	119.2	96.4	2.5	2.00	123.7	1.18	1806	193
33.	Staszic	–	–	95.8	64.3	53.2	35.7	149.0	1.42	1960	50
34.	Szombierki	1.7	0.9	193.4	99.1	–	–	195.1	1.87	1865	125
35.	Śląsk-Matylda	7.2	6.3	106.6	93.7	–	–	113.8	1.09	1857	117
36.	Śląsk	–	–	44.6	64.9	24.2	35.1	68.8	0.66	1974	36
37.	Wawel	6.7	2.8	229.1	96.9	0.8	0.3	236.6	2.26	1752	130
38.	Wesoła	–		161.7	74.8	54.5	25.2	216.2	2.07	1785	96

(continued)

Table 2.2 (continued)

No.	Coal mines (named by the state in 1993)	Coal output								Year of start of mining	Years of activity since the begining of miming statistic
		to 1882[a]		1883–1993		1994–2009		Total	% of output in USCB		
		mln tonnes	%	mln tonnes	%	mln tonnes	%	mln tonnes			
39.	Wieczorek	7.9	3.0	225.6	86.2	28.2	10.8	261.7	2.50	1826	176
40.	Wujek	–	–	144.9	81.7	32.4	18.3	177.4	1.70	1900	110
41.	Ziemowit	–	–	217.8	74.9	73.1	25.1	290.9	2.78	1893	117
Silesian Upland		118.2	1.6	6453.8	87.9	770.9	10.5	7342.9	70.2	1740	240
Racibórz-Oświęcim Basin											
42.	1 Maja	–	–	68.3	95.8	3.0	4.2	71.3	0.68	1953	43
43.	Anna	–	–	128.5	82.2	27.8	17.8	156.2	1.49	1832	127
44.	Borynia	–	–	54.0	59.6	36.7	40.4	90.7	0.87	1971	39
45.	Brzeszcze	–	–	130.6	78.3	36.2	21.7	166.7	1.59	1907	102
46.	Budryk	–	–	–	–	41.6	100.0	41.6	0.40	1994	16
47.	Chwałowice	–	–	96.4	70.0	41.4	30.0	137.7	1.32	1906	104
48.	Czeczott	–	–	19.2	45.7	22.8	54.3	42.0	0.40	1984	19
49.	Dębieńsko[b]	–	–	100.1	90.5	10.5	9.5	110.6	1.06	1899	102
50.	Gliwice	–	–	59.9	90.9	6.0	9.1	65.9	0.63	1911	89
51.	Janina[b]	–	–	91.6	72.5	34.8	27.5	126.4	1.21	1907	98
52.	Jankowice	–	–	128.5	69.2	57.2	30.8	185.7	1.78	1916	94
53.	Jastrzębie	–	–	73.8	62.4	44.4	37.6	118.3	1.13	1963	47
54.	Knurów	–	–	158.6	78.0	44.8	22.0	203.4	1.94	1906	104
55.	Krupiński	–	–	11.1	24.9	33.6	75.1	44.7	0.43	1985	25

(continued)

Table 2.2 (continued)

No.	Coal mines (named by the state in 1993)	Coal output								Year of start of mining	Years of activity since the begining of miming statistic
		to 1882[a]		1883–1993		1994–2009		Total	% of output in USCB		
		mln tonnes	%	mln tonnes	%	mln tonnes	%	mln tonnes			
56.	Makoszowy	–	–	180.3	80.3	44.4	19.7	224.6	2.15	1906	103
57.	Marcel	–	–	123.1	73.3	44.8	26.7	167.8	1.60	1883	127
58.	Morcinek	–	–	5.2	45.3	6.3	54.7	11.5	0.11	1987	12
59.	Moszczenica	–	–	78.7	97.9	1.7	2.1	80.4	0.77	1966	29
60.	Piast[b]	–	–	94.8	55.7	75.4	44.3	170.2	1.63	1976	34
61.	Pniówek	–	–	54.1	49.3	55.6	50.7	109.7	1.05	1974	36
62.	Rydułtowy	1.9	1.0	155.9	81.8	32.9	17.2	190.7	1.82	1792	171
63.	Rymer	–	–	81.3	98.8	1.0	0.2	82.3	0.79	1896	99
64.	Silesia	–	–	47.8	77.2	14.1	22.8	61.9	0.59	1907	97
65.	Sośnica	–	–	157.6	78.6	42.8	21.4	200.4	1.92	1917	93
66.	Szczygłowice	–	–	103.2	70.9	42.4	29.1	145.6	1.39	1961	49
67.	Zofiówka	–	–	66.0	62.4	39.8	37.6	105.8	1.01	1969	41
68	Żory	–	–	10.4	89.8	1.2	10.2	11.6	0.11	1979	18
Racibórz-Oświęcim Basin		1.9	0.06	2278.6	72.94	843.0	27.0	3123.5	29.8	1792	171
Mining area		120.1	1.2	8732.4	83.4	1613.9	15.4	10466.4	100.0	1740	240

[a]Research periods were established according to the editions of topographic maps used in the morphometric analysis
[b]The mine is located on the border of the Silesian Upland and the Racibórz-Oświęcim Basin

In the Upper Silesian Coal Basin, by 2009, nearly 10.7 billion tonnes of hard coal were output, of which up to 71 % after 1950. Waste rock mining is difficult to estimate. Assuming an average production of 0.2 to 0.4 tonnes of rock per 1 tonne of coal produces a number of 2.1–4.3 billion tonnes. The total output of hard coal and waste rock from the bedrock of the USCB amounted to 13–15 billion tonnes. In the three periods, analysed in the following part of the book, the exploitation was as follows: from 1769 to 1882 (i.e. within 114 years), 120.1 million tonnes of coal were extracted, which was 1.2 % of the output up to date (If small coal mines are included, the numbers are 277.2 million tonnes and 2.6 %). Almost all the coal production then came from the area of the Silesian Upland (98.4 %); an insignificant part of the coal was extracted in the Racibórz-Oświęcim Basin. In the period of 1883–1993 (111 years), more than 8.732 billion tonnes of coal were extracted in the research areas, or 83.4 % of total production (If including production from small coal mines, it would be 8.81 billion tonnes, or 82.3 %). Coal mining was still concentrated in the Silesian Upland (73.9 %). In the period of 1994–2009, just over 1.6 billion tonnes of coal were extracted in the USCB (15.4 % of the up to date output). In contrast to previous periods, a majority of production came from the Racibórz-Oświęcim Basin (52 %). In 2014, industrial resources of coal were estimated at 3.44 billion tonnes. In that year, 76.75 million tonnes were extracted (In the period not considered in this work, from 2010 to 2014, a total of 330 million tonnes of coal were extracted).

2.4 Rocks Resources Mining

In the second half of the nineteenth century, the mining of solid rock—limestone and marl—developed in the area of the USCB for the lime and cement industries, dolomite for metallurgy and dimension, and crushed stones (limestone, dolomite, porphyry, melaphyry, and diabase) for various purposes, mostly construction (Fig. 2.3). Mining areas coincide with outcrops of Triassic and Jurassic and locally Permian rocks. Exploitation was concentrated in the south-eastern and northern parts of the USCB. Of the 110 quarries, currently only 8 remain active.

Triassic dolomites occurring in the USCB are characterized by very good quality parameters for metallurgical purposes. For over 115 years (1883–1997), they were exploited in the Bobrowniki-Blachówka and Gródek quarries (Fig. 2.4). Since 1918, dolomite has been extracted from the Żelatowa deposit in Pogorzyce. It is one of four currently active dolomite quarries in Poland; in 2009, 0.676 million tonnes of raw materials were extracted there (23 % of national production).

Triassic limestone mining has developed for purposes of the cement industry from the Sadowa Góra, Żychcice, Rogoźnik, Górka and Płaza deposits (Table 2.3). The last of these deposits has been exploited in two large excavations since 1887, but currently only on a small scale (18,000 tonnes per year). In other quarries, mining was abandoned even though it used to be significant; for example, about 7.2 million tonnes of rocks were extracted from the Sadowa Góra deposit in the years

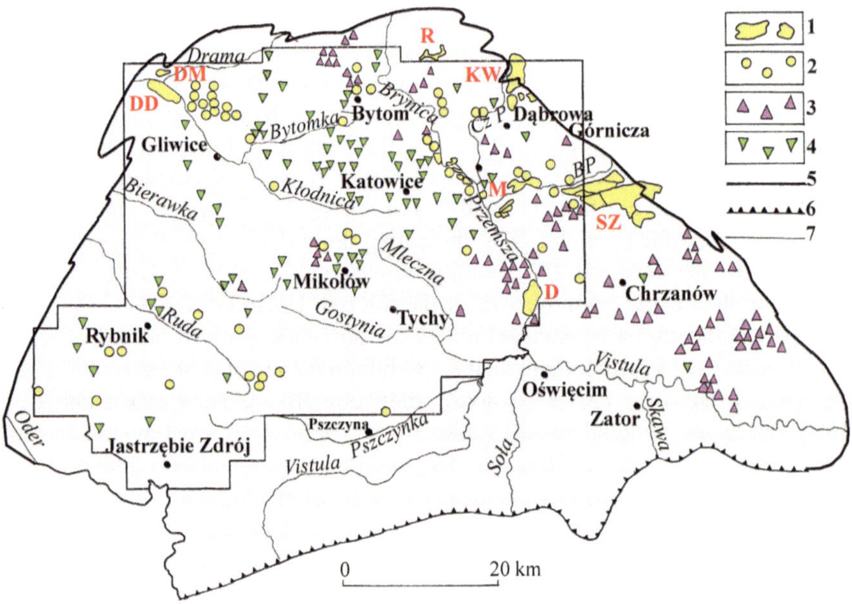

Fig. 2.3 Rocks resources mining in the Upper Silesian Coal Basin (based on Dulias 2013). (**1**) The largest sandpit of stowing sands: *D* Dziećkowice sandpit, *DD* Dzierżno Duże sandpit, *DM* Dzierżno Małe sandpit, *KW* Kuźnica Warężyńska sandpit, *M* Maczki-Bór sandpit, *R* Rogoźnik sandpit, *SZ* Szczakowa sandpit, (**2–4**) excavations of less than 0.6 km^2: (**2**) sandpits, (**3**) quarries (limestones, dolomites, porphyries, melaphyres, diabases), (**4**) clay-pits, (**5**) boundaries of the Upper Silesian Coal Basin after Doktorowicz-Hrebnicki 1968, (**6**) Carpathian thrust, (**7**) the Polish-Czech border; abbreviations: B P—the Biała Przemsza River, Cz P—the Czarna Przemsza River

1954–1980. Limestones and marls for the lime industry were exploited from the Sosnowiec Środula, Brynica-Czeladź and Mikołów-Mokre deposits.

The main areas of rock exploitation for construction purposes are now Imielin Hills (dolomite) and the Tenczynek Hummock (limestones, porphyries, diabases). In the two decades of 1990–2009, the Żelatowa quarry showed the largest output at 17 million tonnes, followed by the Zalas quarry with 14 million tonnes. In each of the remaining quarries, less than 3.8 million tonnes of deposits were extracted (Table 2.3). In addition to large quarries, there are a number of smaller quarries in the area of the USCB, especially in its south-eastern part, which have been used since the nineteenth century for local construction or lime industry needs.

Within the USCB, there are also deposits of gravel aggregates. Operation has been discontinued in 19 and is still carried out in 30. In 2009, a total of 4.776 million tonnes of aggregates were extracted there, which represents 3.4 % of national production. Most of the output came from large river valleys: the Oder (35 %), the Vistula, the Soła, and the Olza (41 %). The total output of gravel aggregates by 2009 was estimated at approximately 300 million tonnes.

Fig. 2.4 Gródek dolomite quarry in the Ciężkowice Hummock (Dulias 2011)

Stowing sands mining has more than a hundred years of history in the USCB (Dulias 2010). Initially, the exploitation of sands from fluvioglacial and partly aeolian deposits was carried out in the vicinity of mines in small, shallow (5–8 m) excavations, up to groundwater level only. Then, until the mid-twentieth century, mining was focused in the Dzierżno Duże, Dzierżno Małe, Pogoria I, Pogoria II, Czechowice, and Betoniarnia sandpits, as well as in the complex of sandpits of the Brynica Valley: Przezchlebie, Borowa Wieś, and Panewniki. Production volumes in the last 3 workings could not be determined because they were filled in with tailings. After World War II, stowing sands were mainly extracted from the Szczakowa, Kuźnica Warężyńska, Dziećkowice, Maczki-Bór Zachód, Maczki-Bór Wschód, Pogoria III, Jęzor Wysoki Brzeg, and Rogoźnik pits. Sandpits occur mainly in the eastern part of the USCB in the valleys of the Biała and Czarna Przemsza, as well as in the valleys of Brynica and Kłodnica.

By 2009, 1.75 billion tonnes of stowing sands were extracted, with more than 93 % from the 6 largest sandpits: Szczakowa, Dzierżno Duże, Kuźnica Warężyńska, Dziećkowice, and Maczki-Bór (Western and Eastern) (Fig. 2.5, Table 2.4).

In the Upper Silesian Coal Basin, there are many excavations that remained after the exploitation of raw clay for construction ceramics. The development of mining of this material was a result of the widespread clay rock—glacial in the west, Miocene in the south, Carboniferous in the central part of the USCB, and Permian

Table 2.3 The main quarries of solid rocks in the Upper Silesian Coal Basin (after Dulias 2013)

Quarry	Location	Type of solid rocks	Period	Output 1990–2009 (mln tonnes)
Dolomites for metallurgical industry				
Bobrowniki-Blachówka	Tarnowice Plateau	Dolomites	Triassic	–
Gródek	Ciężkowice Hummock	Dolomites	Triassic	1.2
Żelatowa	Jaworzno Hummock	Dolomites	Triassic	16.8
Limestones and marls for cement industry				
Sadowa Góra	Jaworzno Hummock	Limestones	Triassic	–
Żychcice	Twardowice Plateau	Limestones	Triassic	–
Rogoźnik	Twardowice Plateau	Limestones	Triassic	–
Płaza	Tenczynek Hummock	Limestones	Triassic	2.48
Dimension and crushed stones				
Libiąż	Libiąż Hills	Dolomites	Triassic	3.2
Imielin North	Imielin Hills	Dolomites	Triassic	2.5
Imielin Rek	Imielin Hills	Dolomites	Triassic	2.6
Imielin	Imielin Hills	Dolomites	Triassic	2.7
Rybna (Balaton)	Myślachowice Hills	Limestones	Jurrasic	–
Pogorzyce	Tenczynek Hummock	Limestones	Triassic	–
Zalas	Tenczynek Hummock	Limestones	Jurrasic	–
Kamień-Odwozy	Tenczynek Hummock	Limestones	Jurrasic	–
Regulice (Czarna Mountain)	Tenczynek Hummock	Melaphyres	Permian	–
Poręba-Żegoty	Tenczynek Hummock	Melaphyres	Permian	–
Rudno-Wymiarki	Tenczynek Hummock	Melaphyres	Permian	–
Zalas	Tenczynek Hummock	Porhyries	Permian	14.5
Orlej in Głuchówki	Tenczynek Hummock	Porhyries	Permian	–

(continued)

Table 2.3 (continued)

Quarry	Location	Type of solid rocks	Period	Output 1990–2009 (mln tonnes)
Miękinia	Myślachowice Hills	Porhyries	Permian	–
Kowalska Mountain	Myślachowice Hills	Porphyritic tuff	Permian	–
Niedźwiedzia Mountain	Tenczynek Hummock	Diabases	Permian	3.8

Fig. 2.5 Maczki Bór sandpit in the Biskupi Bór Basin (Dulias 2008)

in the east. The exploitation, especially in the nineteenth and the first half of the twentieth century, was carried out for numerous brickyards, predominantly of local importance. The output of these clay pits has not been included in official statistics. Of the 46 deposits of clay raw materials, currently only two are in operation. Given the volume of excavations and density of the extracted raw materials, it was estimated that throughout the USCB, about 13 million tonnes of raw clay were extracted.

Table 2.4 Output of stowing sands in the Upper Silesian Coal Basin to 2009 (from Dulias 2013)

Sandpit	Output		% of total output
	(mln tonnes)	(mln m^3)	
Szczakowa	1107.7	651.6	63.13
Dzierżno Duże	188.7	111.0	10.75
Kuźnica Warężyńska	139.6	82.1	7.95
Dziećkowice	109.8	64.6	6.26
Maczki Bór (Western and Eastern)	88.9	52.3	5.07
Pogoria III	30.1	17.7	1.71
Dzierżno Małe	23.8	14.0	1.36
Rogoźnik	12.9	7.6	0.74
Pogoria I	10.5	6.2	0.60
Pogoria II	6.6	3.9	0.38
Milowice	4.4	2.6	0.25
Morawa	3.7	2.2	0.21
Hubertus I–IV	3.1	1.8	0.17
Betoniarnia	2.9	1.7	0.16
Jęzor–Wysoki Brzeg	2.7	1.6	0.16
Rozkówka	1.9	1.1	0.11
Stary Czekaj	1.5	0.9	0.09
Borki Duże	1.2	0.7	0.07
Other sandpits	14.6	8.6	0.83
Total	1754.7	1032.2	100.00

References

Bąk Z, Barańczuk T (1989) Problemy eksploatacji pokładów węgla kamiennego pod zrobami rudnymi likwidowanej kopalni "Orzeł Biały". Ochrona Terenów Górniczych 89(90):23–27

Cabała J, Sutkowska K (2006) Wpływ dawnej eksploatacji i przeróbki rud Zn-Pb na skład mineralny gleb industrialnych, rejon Olkusza i Jaworzna. Prace Nauk Inst Górn Polit Wroc 117 (32):13–22

Czermiński J (ed) (1992) Hutnictwo na ziemiach polskich. Stow Inż Tech Przem Hutn w Polsce, Kraków

Dulias R (2010) Landscape planning in areas of sand extraction in the Silesian Upland. Poland. Land Urban Plan 95(3):91–104

Dulias R (2013) Denudacja antropogeniczna na obszarach górniczych na przykładzie Górnośląskiego Zagłębia Węglowego. Wyd Uniw Śląskiego, Katowice

Górnictwo węgla kamiennego w 40-leciu Polski Ludowej. Szkic monograficzny. GIG, Katowice, 1988

Jaros J (1969) Historia górnictwa węglowego w Zagłębiu Górnośląskim w latach 1914–1945. Śląski Inst Nauk, Katowice-Kraków

Jaros J (1973) Historia górnictwa węglowego w Polsce Ludowej (1945–1970). Śląski Inst Nauk, PWN, Warszawa-Kraków

Jaros J (1975) Zarys dziejów górnictwa węgla. Warszawa

Karbownik A, Wodarski K (2005) Efekty restrukturyzacji polskiego górnictwa węgla kamiennego w latach 1990–2004. Wiad Górn 6:274–281

Kossuth S (1961) Zarys rozwoju techniki górniczej w kopalniach węgla w Zagłębiu Górnośląskim do połowy XIX wieku. Wyd Geol, Warszawa

Kossuth S (1965) Górnictwo węglowe na Górnym Śląsku w połowie XIX wieku. Ślask, Katowice, Prace GIG, Wyd

Kossuth S (1968) Zarys historyczny rozwoju metod eksploatacji w polskim górnictwie węglowym do 1914 roku. In: Borecki M (ed) Monografia polskiego górnictwa. Systemy eksploatacji węgla, GIG

Kotyrba A (2005) Zagrożenie i ryzyko zapadliskowe terenów Górnośląskiego Zagłębia Węglowego. Wiad Górn 7–8:348–358

Kowalski A (1996) Wpływ eksploatacji na deformacje powierzchni na przykładzie eksploatacji KWK Staszic. Mat Konf Szk Ekspl Podz, Katowice, pp 139–152

Krupiński B (1956) Analiza eksploatacji filaru ochronnego pod miastem. Drogi postępu w górnictwie 1. Warszawa: 239–319

Lamparska-Wieland M (2003) Atlas zmian wybranych elementów krajobrazu terenów górniczych i pogórniczych Płaskowyżu Tarnowickiego. Prace WNoZ UŚ 27, Warszawa

Luksa J (1959) Rozwój wydobycia w kopalniach węgla kamiennego w Polsce w latach 1769-1948. Studia i Materiały PTE, Katowice

Messtischblätter, Herausgegeben von der Preußischen Landesaufnahme, 1:25 000 Reichsamt für Landesaufnahme, Berlin 1883

Minorczyk R (1986) 125 lat Zakładów Górniczo-Hutniczych „Orzeł Biały". Rudy Met Nieżel 31 (12):462–469

Molenda D (1972) Kopalnie rud ołowiu na terenie złóż śląsko-krakowskich w XVI-XVIII wieku. Inst HistKultury Materialnej PAN. Zakł Narodowy im. Ossolińskich

Musioł L, Płuszczewski S (1960) Wykaz zakładów dawnego hutnictwa żelaza na Górnym Śląsku od XIV do połowy XIX wieku. Studia z dziejów górnictwa i hutnictwa 5. Warszawa-Wrocław: 7–87

Nowak J (1927) Kronika miasta i powiatu Tarnowskie Góry. Najstarsze dzieje Śląska i Ziemi Bytomsko-Tarnogórskiej. Dzieje pierwszego górnictwa w Polsce, Tarnowskie Góry

Piernikarczyk J (1933/1934) Historia górnictwa i hutnictwa na Górnym Śląsku. T. 1, 2. Śląski Zw Akad, Katowice

Popiołek F (1965) Górnośląski przemysł górniczo-hutniczy w drugiej połowie XIX wieku. Śląski Inst Nauk, Katowice

Radwan M (1963) Rudy, kuźnice i huty żelaza w Polsce. Wyd Nauk Techn, Warszawa

Rozmus D, Bodnar R (2004) Wczesnośredniowieczne ślady hutnictwa metali nieżelaznych w Dąbrowie Górniczej – Łośniu oraz na obszarach przygranicznych. In: Rozmus A (ed) Archeologiczne i historyczne ślady górnictwa i hutnictwa na terenie Dąbrowy Górniczej i okolic. Księgarnia Akad, Kraków, pp 9–60

Skinderowicz B (1963) Eksploatacja filarów ochronnych w górnictwie węglowym. Przegl Techn 4

Tkocz M (1987) Koncentracja górnictwa węgla kamiennego w województwie katowickim. In Problemy geograficzne górnośląsko-ostrawskiego regionu przemysłowego. ODN IKN, WNoZ UŚ Katowice - Sosnowiec: 106–110

Żeglicki J (1996) O budowie geologicznej, rudach i minerałach rejonu Tarnowskich Gór. Stow. Miłośn Ziemi Tarnogórskiej, Tarnowskie Góry

Ziemba J (1967) Biedaszyby Górnego Śląska i Zagłębia Dąbrowskiego. Wyd Śląsk, Katowice

Chapter 3
Anthropogenic Landforms in the Upper Silesian Coal Basin

At the end of the nineteenth century, the landscape of the Upper Silesian Coal Basin was similar to the landscapes of other European coal basins of the Industrial Revolution. Mine shafts, factory and ironworks chimneys, heaps, and pits were all shrouded with smoke and covered with black dust. Features of the landscape at the time, as well as its mining identity, were very well reflected in the term *Czarny Śląsk* (Black Silesia), which stuck to the region for nearly a century. The scale of environmental degradation was noticed in the postwar period. The problem even became a matter of national importance: a resolution of the Presidium of the Council of Ministers of 6 June 1953 imposed on the Polish Academy of Sciences the obligation to conduct scientific research in the area of the Upper Silesian Industrial Region (Greszta 1957). The results of the research conducted within the framework of numerous committees—including ones devoted to the matter of bedrock, development of post-industrial brownfields (later, soil sciences and mining), and utilizing heaps—were published in a series of newsletters devoted to the USIR. A synthetic diagnosis of the relief transformation in the middle of the twentieth century was the USIR geomorphological map on a scale of 1: 50,000 (Mapa geomorfologiczna Górnośląskiego Okręgu Przemysłowego 1959). In the last half-century, a number of works have been written on the direct and indirect human impact on the geomorphology of areas located within the Upper Silesian Coal Basin (e.g. Karaś-Brzozowska 1960; Hornig 1968; Żmuda 1973; Jania 1983; Jankowski 1986; Szczypek and Wach 1991a; Helios-Rybicka and Rybicki 2002; Rzętała M.A. 2003; Kupka et al. 2005; Pełka-Gościniak 2006; Wojciechowski 2007; Dulias 2010, 2011; Solarski and Pradela 2010). They describe various aspects of the anthropogenic transformation of the relief, developed with the use of different methods and with different degrees of detail. Mainly, the work by Żmuda (1973) had regional character: it was the first comprehensive depiction of environmental changes in the area of the Upper Silesian conurbation. More recently, a work by Dulias (2013) was published on the anthropogenic denudation of mining areas of the USCB. A particularly valuable cartographic work was the map of relief transformations of the Katowice Voivodeship on a scale of 1: 50,000, developed

R. Dulias, *The Impact of Mining on the Landscape*,
Environmental Science and Engineering, DOI 10.1007/978-3-319-29541-1_3

Table 3.1 Anthropogenic landforms caused by direct mining activity in the Upper Silesian Coal Basin (from Dulias 2013)

Anthropogenic landforms	Number of landforms	Area (km^2)	Volume (mln m^3)
Spoil tips	302	49.2	370.0
Quarries	110	6.1	97.7
Clay pits	115	3.0	6.5
Sandpits	109	76.3	1032.2
Gravel aggregate pits	49	15.0	147.0
Total	685	149.6	1653.4

with a photo-interpretation method, which represented anthropogenic landforms and isolines of surface subsidence (Mapa przeobrażeń powierzchni ziemi 1982).

Mining in the Upper Silesian Coal Basin has contributed to the formation of direct and indirect anthropogenic relief forms (Table 3.1). The most outstanding and most easily recognizable elements of landscape are the forms related to direct mining operations: many of them have retained their morphological distinctness for tens or even hundreds of years. Post-mining relief forms have frequently been transformed deliberately: radical changes in the dimensions and shape occurred, particularly in the case of filling in excavations with mining waste and forming convex dumps in their place. Direct anthropogenic relief forms occupy an area of less than 150 km^2. Concave forms outweigh convex forms, both in terms of volume (56 %), and surface (67.1 %), as well as cubic capacity (77.6 %). The most important role in the anthropogenic relief of the USCB is played by sandpits, occupying more than half of the surface of all post-mining forms and over 62 % of their volume. In many sandpits, there are water reservoirs; this is why these forms, despite their anthropogenic origins, are well integrated into the landscape (Dulias 2010). Spoil tips constitute a strong accent in the relief of the USCB.

The occurrence of direct anthropogenic landforms in individual geomorphological mesoregions is varied. Their largest percentage is in the North Silesian Upland at almost 14 %, with primarily concave forms. This mesoregion covers over 38 % of area and 51 % of all anthropogenic forms in the USCB. Convex forms (spoil tips) occupy the largest area in the Southern Silesian Upland. On the scale of lower-order geomorphological units, direct anthropogenic landforms have the largest total area in the Biskupi Bór Basin at 38.3 km^2; the Czarna Przemsza Valley has 8.5 km^2 and the Mysłowice Basin has 7.9 km^2. They take up such a large part of the surface of the said units (28.5, 31.9, and 13.8 % respectively) that the relief in these areas has an exceptionally anthropogenic character. Spoil tips dominate in the anthropogenic relief of the Ruda Hills (6.2 % of the area) and the Siemianowice Upland (5.6 %).

In the remaining geomorphological units, direct anthropogenic landforms occupy about 1 % of the area. Indirect anthropogenic landforms include forms that have been created as a result of continuous and discontinuous surface deformations induced by geomechanical rock mass transformation in underground mining areas. These include subsidence troughs, sinkholes, fissures, and thresholds.

3.1 Excavations

Post-mine holes. The oldest mining landforms are shallow holes that remained on the surface due to the former surface or small shaft operation of different ores. They are accompanied by mounds and heaps formed from the removed overburden and waste rock, which were called *warpie* (Fig. 3.1). The dimensions of the holes and mounds are small; they usually are several meters in diameter and are of similar depth or height. These forms are found in larger clusters, mostly on hummock and Triassic hill culminations, where shallow exploitation of ores was carried out on outcrops up to the groundwater level. The younger the forms, the clearer are their morphological contours. However, even centuries-old forms are recognizable in the terrain. The resulting unevenness of the substrate prevented subsequent use; for example, in agriculture these areas became overgrown, which contributed to maintaining their forms. Larger concentrations of post-mine holes are present on the Tarnowice Plateau, the Ząbkowice Hummock, and the Jaworzno Hummock (Fig. 3.2). On the Ojców Plateau, in the vicinity of Płoki, in an area of 8 km^2, over 400 pits have been accounted for, with a diameter of 20 m and at a depth of up to 7 m (Górecki and Szwed 2005).

Quarries are found in the northern and eastern part of the Upper Silesian Coal Basin. They are mostly about 100–150 years-old forms; however, there are also older examples, such as the porphyry quarry in Miękinia in Myślachowice Hills, which has operated since the seventeenth century. Quarries generally occupy a

Fig. 3.1 Old post-mine holes and mounds (*warpie*) in the Ząbkowice Hummock (Dulias 2004)

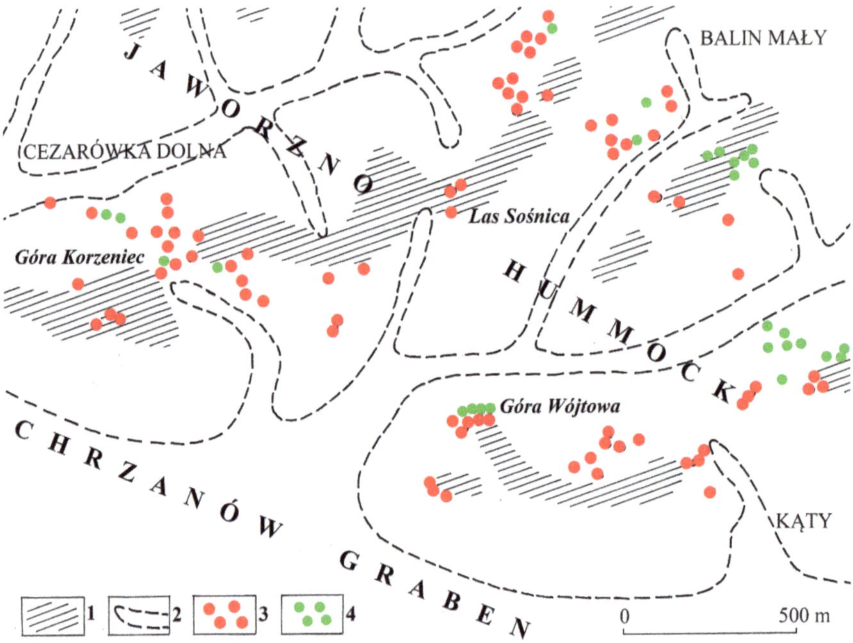

Fig. 3.2 Old post-mine landforms after surface exploitation of ores in the Jaworzno Hummock (from Dulias 2013). (**1**) Flat-topped hills, (**2**) dry erosional-denudational valleys, (**3**) old post-mine holes, (**4**) small mounds, embankments (*warpie*)

small area: out of the 110 inventoried forms, only 12 have an area greater than 0.2 km^2. The largest are the dolomite quarries of Żelatowa (0.59 km^2) and Bobrowniki-Blachówka (0.55 km^2). Quarries have from several to several dozen meters of depth and can be multilevel (Fig. 3.3). The height of the exploitation walls of each level is 6–15 m (rarely more than 20 m). The walls are steep, with different stability. Most of them are subject to weathering and rock mass movement (Fig. 3.4).

Subsurface quarries are mostly situated on flattened hill culminations (e.g. the melaphyres quarry from the Poręba-Żegota deposit), while slope quarries cut into

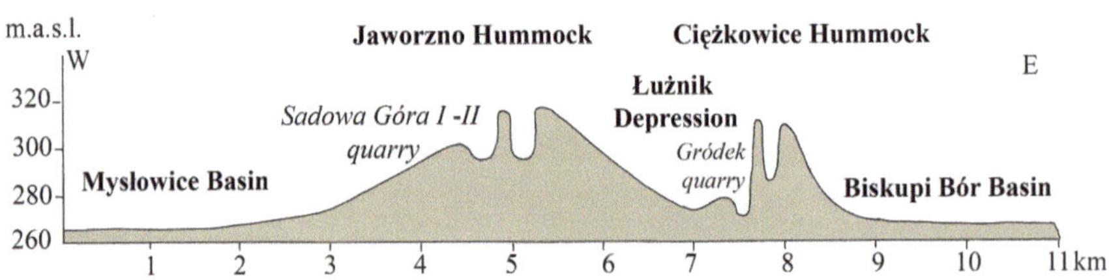

Fig. 3.3 Morphological cross-section through quarries on the Jaworzno and Ciężkowice Hummocks (from Dulias 2013)

Fig. 3.4 Dolomite quarry in the Ciężkowice Hummock (Dulias 2014)

the slopes of valleys or hills (e.g. the Żelatowa dolomite quarry). The intensive exploitation of hard rock in some areas has led to an inversion of relief. An example is the quarry of Jurassic limestone from the Rybna deposits (Myślachowice Hills), established on a hill that is 346 m above sea level; currently, it is a depressed area with a water reservoir, whose water surface is located at an altitude of 318 m above sea level (Głogowska 2007).

Most quarries (29) are found on the Tarnowice Plateau; however, apart from a few large pits, most of them constitute small forms. The volume of quarries in this geomorphological unit represents only 6.5 % of the total volume of the quarries in the USCB. Those in the Tenczynek Hummock represent up to 39 % (16 quarries), the Jaworzno Hummock has 20 % (8 forms), and Myślachowice Hills has more than 18 % (4 forms). The quarries occupy the largest area on the Tenczyn Hummock (1.5 km^2); the Tarnowice Plateau and the Jaworzno Hummock are 0.9 km^2. The area of quarries totals just over 6 km^2. Their volume is estimated at 97.7 million m^3. The biggest excavations in terms of volume are Żelatowa (18 million m^3), Zalas, Miękinia, Plaża, Niedźwiedzia Góra, Blachówka-Bobrowniki, Regulice, Żychcice and Rybna.

Openpit mines. In the USCB, there were many open-pit mines, mainly coal. In fewer numbers, there were also zinc, lead, and iron mines, but only a few of them were large. The biggest coal open pits were Reden and Koszelew on the Dańdówka Plateau and Paryż and Brzozowica in the Dąbrowa Basin (Fig. 3.5). In each of them, the 510 seam was excavated, which was the thickest coal seam in the Upper Silesian Coal Basin (up to 24 m). The Reden open pit had a length of over 1 km

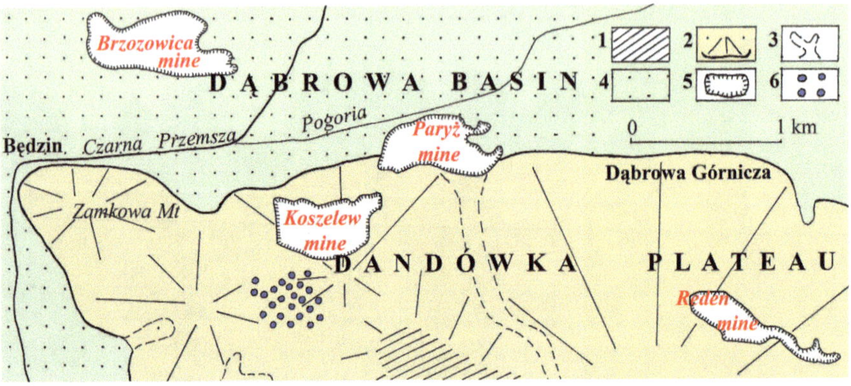

Fig. 3.5 Open-pit coal mines against a background of the geomorphological sketch of the borderline between the Dąbrowa Basin and the Dańdówka Plateau, in the mid-twentieth century (from Dulias 2013). (**1**) Flat-topped hills, (**2**) slopes, (**3**) dry erosional-denudational valleys, (**4**) river valleys, (**5**) open-pit coal mines, (**6**) old post-mine holes

and a width of 200 m; excavation slopes, despite the partial backfilling, are still visible in the relief. Starościak (2006) quoted a description of an open pit of 1843 "… a huge pit, dug vertically in the ground, a few hundred steps wide here and there, a few dozen ells deep and nearly a mile long, extending in different bent and broken directions." In 1956, the Brzozowica open pit was established within the limits of the Paryż mine; it operated until 1968 (Ciepiela 2003). The total area of the mine was 0.5 km^2, with a depth starting from the deck floor of 510, 35–40 m in the north and 90 m in the south (Rechowicz 1974). The open pit, with a capacity of 13 million m^3, was completely covered with ashes from power plants, similarly to the >50 m-deep Paryż pit.

Open-pit coal mines were also established on Carboniferous outcrops of the Murcki Plateau and the Mikołów Hummock and in the Basins Mysłowice and Biskupi Bór, where Carboniferous rocks are under the cover of Quaternary sediments. Out of the total number of 37 major open-pit coal mines, as many as 33 were formed in the 1950s and 1960s. The oldest open-pit mines are Reden (1785) and Koszelew (1825). Ore open pits were located in the Bytom Plateau (Szarlej) and in the eastern part of the USCB, in the Wilkoszyn Syncline (Balin).

Sandpits. In the USCB, there are 109 pits where sand used to be exploited, mostly stowing sand. They cover a total area of 76.3 km^2, of which over 42 % is attributable to 5 mine workings of the Szczakowa mine and 33 % to the following 5 large sandpits: Dziećkowice, Kuźnica Warężynska, Dzierżno Duże, Maczki-Bór Zachód, and Maczki-Bór Wschód. The largest area is occupied by sandpits in the eastern part of the USCB in the area of Biskupi Bór, Dąbrowa and Mysłowice Basins, and the Chrzanów Graben—a total of 62.5 km^2 (82 %). The volume of sandpits exceeds 1 billion m^3. In the mining period, sandpits are generally of geometric shapes, with steep slopes limiting each exploitation level (2–4) and a flat bottom. The depth of large excavations is, on average, 15–20 m, with a maximum of about 30 m (Dulias 2010). In sandpits, fairly intensive aeolian processes take place (Szczypek and Wach 1991a, b, 1993, 1999).

In many sandpits, coal mining waste was stored and eventually turned into above-ground spoil tips, such as in Brzezinka, Przezchlebie, Panewniki, and others. Currently, the two excavations of the Maczki-Bór sandpit are being backfilled. In many sandpits, including the largest ones, there are water reservoirs; their edges are undergoing shoreline processes (Rzętała M.A. 2003; Machowski et al. 2006; Rzętała 2008; Machowski 2010). Old, waterless sandpits are overgrown, but still have distinct edges and steep slopes.

Clay pits. Workings that followed the exploitation of raw clay materials, commonly called clay pits, are numerous (115) but highly dispersed. Most of them are located in the Miechowice Upland (15), the Mikołów Hummock (11), and the Rachowice Plateau (8). Generally, these are small forms: only a few of them have an area greater than 0.05 km^2; an average area is 0.03 km^2. Clay pits in the USCB total only 3 km^2 and their depths are varied—from a few to several meters, rarely more than 20 m. The volume of clay pits totals 6.5 million m^3. Their slopes are steep, actively modelled by denudation processes and water erosion. The bottoms of most excavations contain water reservoirs. The largest forms include the clay pits in Turzyczka on the Rybnik Plateau (0.15 km^2) and in Katowice-Giszowiec and Brynów on the Murcki Plateau (0.1 km^2). In terms of volume, the biggest excavations are located on the Murcki Plateau: the volume of the 5 clay pits present here is 19.2 % of the total volume of these forms in the USCB. Further positions are taken by the Rybnik Plateau (8.9 %) and the Mikołów Hummock (8.2 %). The morphological importance of clay pits in the landscape is durable but insignificant.

3.2 Spoil Tips

Spoil tips are anthropogenic forms that are particularly highlighted in the landscape of the Upper Silesian Coal Basin. These are mainly heaps of waste rock and tailings associated with the mining of coal, zinc, and lead ores and, to a lesser extent, rock materials. The mine tailings represent about 80 % of all waste that was accumulated in heaps of the USCB until 1993; the rest consists of waste from power plants and ironworks. The analysis includes all spoil tips visible on topographic maps of 1: 10,000 from 1993 and of an area larger than 0.5 hectares, regardless of the origin of the waste and the management method (open, closed, or reclaimed).

The number of spoil tips in the area of the USCB in 1993 amounted to 302, covering a total area of 49.2 km^2, of which 110 spoil tips were small or very small, with a capacity of less than 50,000 m^3. There are 192 that are clearly highlighted in the relief (Fig. 3.6). Most heaps occurred in the area of the Ruda Hills and the Rybnik Plateau (39 in each), in the Siemianowice Upland (29) and the Mysłowice Basin (19). These forms covered the largest area in the Rybnik Plateau at nearly 9 km^2, as well as in the Mysłowice Basin (5.4 km^2) and the Miechowice Upland (5.3 km^2). Almost a quarter of the total volume of spoil tips were found in the

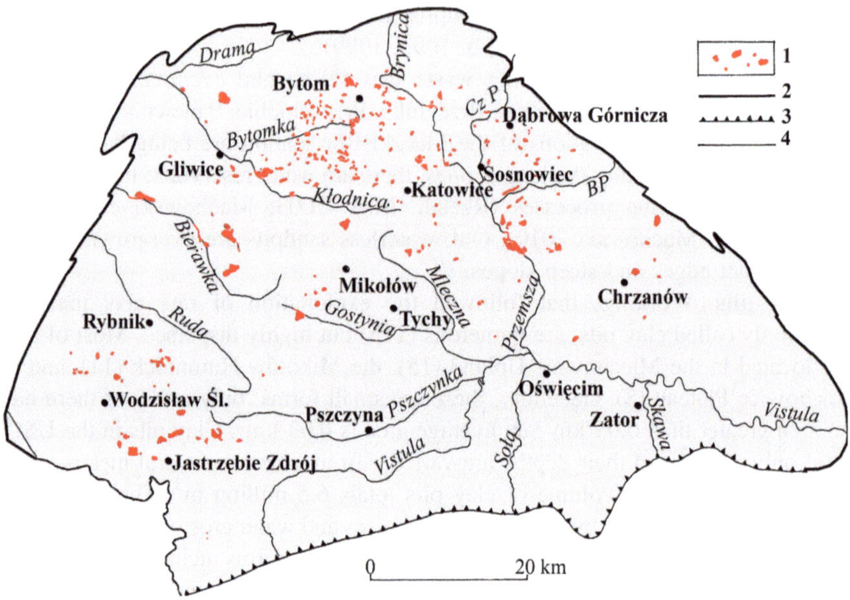

Fig. 3.6 Location of mining spoil tips in the Upper Silesian Coal Basin (from Dulias 2013). (**1**) Spoil tips, (**2**) boundaries of the Upper Silesian Coal Basin after Doktorowicz-Hrebnicki 1968, (**3**) Carpathian thrust, (**4**) the Polish-Czech border. Abbreviations: *BP* the Biała Przemsza River, *CzP* the Czarna Przemsza River

Rybnik Plateau, which is a geomorphological unit where heaps remarkably stand out in the landscape (the largest number, area, and volume of forms). The sizes of the largest spoil tips in the area of the Upper Silesian Coal Basin are comparable to some natural elements of relief, especially with monadnocks (Fig. 3.7)

Spoil tips take different shapes. The older forms are most often conical, such as the characteristic heaps of the Dębieńsko mine in the Southern Podstokowa Zone (Fig. 3.8). Many spoil tips are in the shape of massive mesas, such as the great waste heap of the Bolesław Śmiały coal mine on the Mikołów Hummock (Lamparska-Wieland and Waga 2002) (Fig. 3.9). A lot of heaps are irregular, with several culminations. Some forms have impressive heights: the highest is considered to be the Szarlota (Charlotte) on the Rybnik Plateau, which, depending on the source, is from 80 m in 1986 (Pełka and Pociecha 1991) to 134 m in 2001 (Gawor and Szmatłoch 2010); according to this study, it is 101 m high (1993). The highest spoil tips are on the Rybnik Plateau: among 8 forms that are higher than 65 m, as many as 5 are located within the boundaries of this geomorphological unit, including 3 in the area of the Rydułtowy mine. One of the spoil tips, Dębieńsko, and the previously mentioned Bolesław Śmiały mine waste heap, are almost 90 m high; the spoil tips of the Borynia and Marcel mines reach about 70 m. There were 37 more heaps of a height above 20 m and as many as 113 under 5 m.

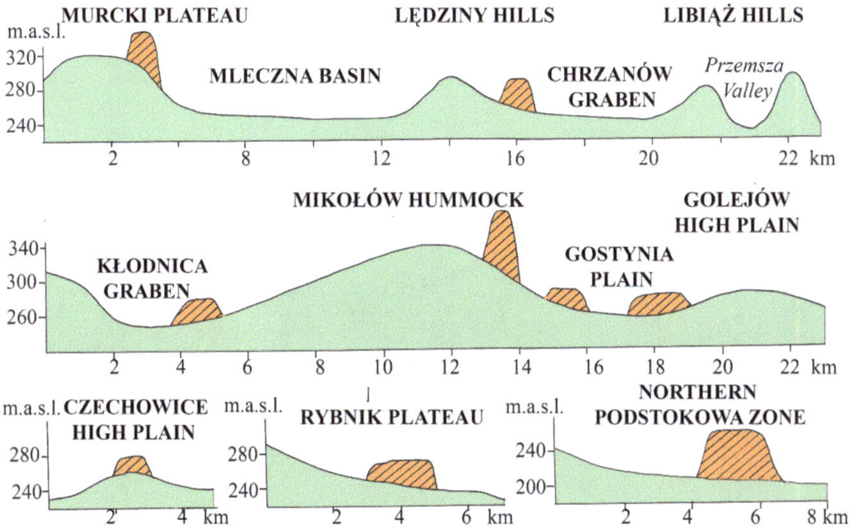

Fig. 3.7 Location and size of the largest spoil tips in relation (1:1) to the natural landforms (based on Dulias 2013)

Fig. 3.8 Old conical spoil tips in Dębieńsko, the Podstokowa Zone (Dulias 2005)

Fig. 3.9 Massive spoil tip in the area of Bolesław Śmiały mine, the Mikołów Hummock (Dulias 2005)

Many spoil tips occupy a very large area: 20 forms of the basis exceeding 0.5 km^2 were accounted for, including 9 with an area of more than 1 km^2. The largest spoil tips of this regard included the heaps of the Knurów mine (1.7 km^2) and the Sośnica mine (1.5 km^2), as well as heaps located in non-mining areas such as the Mysłowice Basin (1.6 km^2), the Czechowice High Plain (1.5 km^2), and the Gostynia Plain (1 km^2). Numerous, extensive heaps are located on the Rybnik Plateau.

In terms of volume, the biggest spoil tips in the USCB are located in the area of the Sośnica mine (25 million m^3), Borynia (22.8 million m^3), Dębieńsko (13.8 million m^3), Marcel (13.4 million m^3), and the non-mining area within the Czechowice High Plain (13.5 million m^3). The greatest volume, however, has the post-flotation settling tank in the zinc and lead ore mine Trzebionka, located in the Krzeszowice Graben (about 27.6 million m^3). Among all 302 spoil tips, 6 of them accumulate one-third of the total volume. In all, 47 forms have a volume of more than 1 million m^3.

Based on a morphometric analysis, the volume of waste that accumulated on the spoil tips of the Upper Silesian Coal Basin was calculated at 370 million m^3 (Table 3.2). Considering the fact that 80 % of the heaps are made of mining waste and assuming that 1 tonne of such waste has a volume of 0.38 m^3 (Żmuda 1973), it can be estimated that about 974 million tonnes of waste is accumulated in the spoil tips. According to statistical data for 1994 in the former province of Katowice (where, de facto, most heaps are located), more than 795 million tonnes of waste accumulated; however, the data do not include a number of minor forms included in morphometric analysis. Both values differ greatly from waste rock extraction in coal

Table 3.2 Spoil tips and excavations formed until 1993 in the geomorphological mesoregions of the Upper Silesian Coal Basin (from Dulias 2013)

Geomorphological mesoregions	Spoil tips		Excavations		Direct anthropogenic landforms	
	Area (km^2)	Volume (mln m^3)	Area (km^2)	Volume (mln m^3)	Area (km^2)	% of mesoregion area
Northern Silesian Upland	4.0	28.5	53.1	814.8	57.1	13.9
Southern Silesian Upland	23.5	128.4	19.1	132.2	42.6	3.6
Racibórz Basin	10.4	84.2	15.5	183.6	25.9	3.7
Oświęcim Basin	10.5	100.0	10.1	92.9	20.6	0.9
Northern Cracow Upland	0.1	0.8	0.9	19.0	1.0	1.6
Southern Cracow Upland	0.7	28.1	1.7	40.9	2.4	0.7
Upper Silesian Coal Basin	49.2	370.0	100.4	1283.4	149.6	3.3

mining, which is estimated, depending on the assumed index, at 2.1–4.3 billion tonnes. This means that at least half of the produced waste was disposed of by means other than storage in spoil tips; for example, it was used for ground levelling. Subsurface dumping that fills subsidence troughs to level them with the surface was carried out in an area up to 60 km^2. A substantial part of mining waste also was used to fill old sandpits, for in roadwork, and for the levelling of land for construction.

Since 1968, a part of mining waste has been used to build embankments along riverbeds—the so-called mine tailings embankments, which are created in valleys with surface subsidence in order to prevent flooding of adjacent areas. They have been built mainly in river valleys—the Kłodnica, the Bierawka, the Szotkowka, the Vistula, the Pszczynka, the Goławiecki Stream, the Czarna Przemsza, the Brynica, the Szarlejka, the Biała Przemsza, and the Bobrek (Skarżyńska et al. 1988). Until the 1990s, about 5 million m^3 of waste was used for the embankment of riverbeds (Pietrzyk-Sokulska 1995).

3.3 Subsidence Troughs

Continuous surface deformations caused by the underground exploitation of raw materials are present in all Polish coal basins as well as in areas of copper, sulphur, and rock salt exploitation. However, nowhere did they reach such large volumes as in the Upper Silesian Coal Basin. It primarily happened due to the enormous scale of underground mining. Until 2009, 13–15 billion tonnes of coal, ore, and waste

Table 3.3 Geological and mining conditions of exploitation and the size of subsidence in Ruda Śląska until 1991, the Ruda Hills (from Dulias 2013)

City district	Thickness of exploited seams (m)	Depth of deposits (m)	Years of exploitation	Maximum subsidence (m)
Wirek	45	5–900	1828–1991	9
Ruda	40	50–580	1750–1991	28
Chebzie	35	40–600	1850–1991	24
Nowy Bytom	33	10–600	1835–1991	18
Godula	32	60–580	1856–1983	24
Bielszowice	29	5–970	1803–1991	14
Bykowina	28	15–700	1870–1991	16
Kochłowice	25	12–970	1824–1991	11
Orzegów	23	60–550	1857–1976	16
Halemba	18	170–970	1921–1991	13

rock were extracted here, running operations mostly by the roof-collapsing method as well as a substantial total thickness of coal seams. In central and north-western parts of the USCB, they even reached 40–60 m, such as in the area of Ruda Śląska (Table 3.3).

The first subsidence troughs appeared in the relief at the beginning of the twentieth century, when coal production increased and exploitation reached significant depths. The period of greatest mining subsidence, however, was observed in the 1960–1980s—three decades of very intensive coal mining. The technology of operations used in the USCB resulted in subsidence as high as 40–60 mm per day, such as in the mines of Sosnowiec, Murcki, and the Rybnik Coal Area (RCA) (Skinderowicz 1982; Kowalski 1996; Białek and Mielimąka 1999). Forecasts for mining subsidence, developed still in the socialist economy, expected a significant intensification of this process in the 1990s. For example, in the RCA, the target subsidence was to exceed 33–35 m; in the area of Knurów and Zabrze, it was to exceed 30 m (Jankowski 1986; Wach 1991). The economic crisis at the end of the last century, the following liquidation of many mines, and a reduction in the volume of mining partly limited the scale of mining subsidence (Fig. 3.10).

Regardless of the surface lowering caused by the extraction of hard coal in the USCB area, they are also associated with natural neotectonic movements. In 1964, Kowalczyk showed that the southern part of the Upper Silesian Coal Basin (subsidence of benchmarks beyond the effects of mining) had an average annual reduction of −0.5 to −1.5 mm in the Vistula basin and from −1 to −3 mm in the Oder basin. This has been confirmed by contemporary research: the south-western part of the USCB, mainly the upper Oder valley, has undergone some of the largest neotectonic movements in Poland, exceeding even 3 mm/year (Zuchiewicz 2000). It must be assumed that the neotectonic movements also affected the areas under mining impact. In the analysed mining period (1883–1993) over a century long, the lowering of the surface resulting from these movements could range from 5.6 cm to a maximum of 38.9 cm.

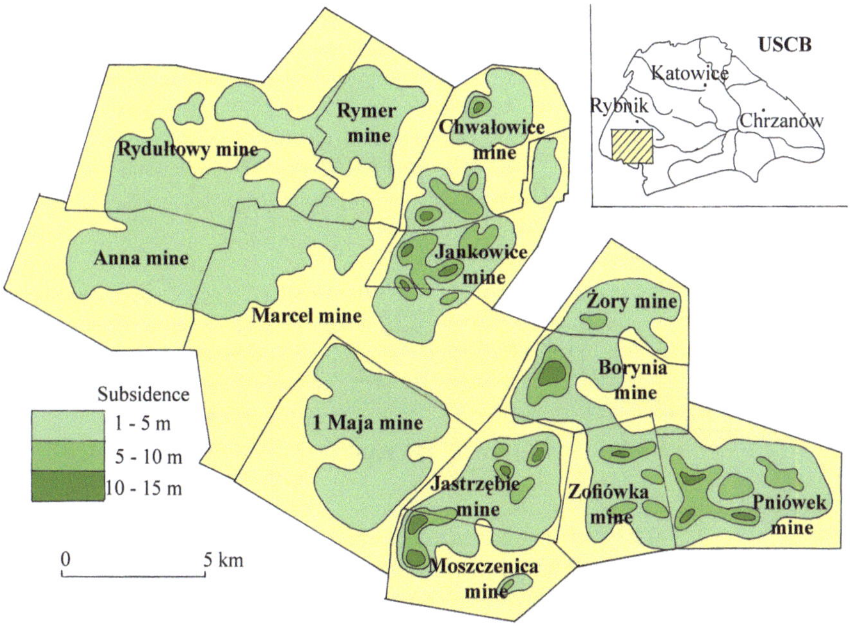

Fig. 3.10 Subsidence in the mining area of the Rybnik Plateau at the turn of the 1970s and 1980s (based on Mapa przeobrażeń powierzchni ziemi woj. katowickiego 1982; Jankowski 1986)

Studies of mining surface deformations are carried out with geodetic methods, including GPS methods. Non-geodetic methods are used less frequently, such as photogrammetric (due to the low precision in comparison with levelling) and laser methods (because of the cost, despite their high accuracy). A method used since the 1990s is satellite radar interferometry (InSAR), which allows for the tracking of vertical changes in relief with centimetre accuracy; for the USCB area, it was used by Perski (2000) and Perski and Jura (1999). Maps forecasting surface subsidence due to underground mining have been developed for individual mines. It should be noted that the forecast subsidence and actual subsidence often differ significantly (e.g. Popiołek and Ostrowski 1981; Hejmanowski and Malinowska 2009). For example, significant differences between the theoretical and the actual course of the final curvature forecast by computer programmes were revealed in the Budryk mine, based on geodetic observation (Mielimąka 2006).

The range of continuous deformations in the USCB in the early 1990s is presented on a map developed by Perski (2000). This map also marks areas of subsidence with rates greater than 10 mm per month, as observed in October 1992 using SAR interferometry (Fig. 3.11). The mining impact area covered 1,185 km^2, including the area adopted in this study −1.124 km^2. This shows a result almost identical to the one obtained by the author −1125.4 km^2.

The range and size of surface lowering were obtained from the subtraction of digital elevation models (DEM) for 1883 and 1993. It was assumed that the model

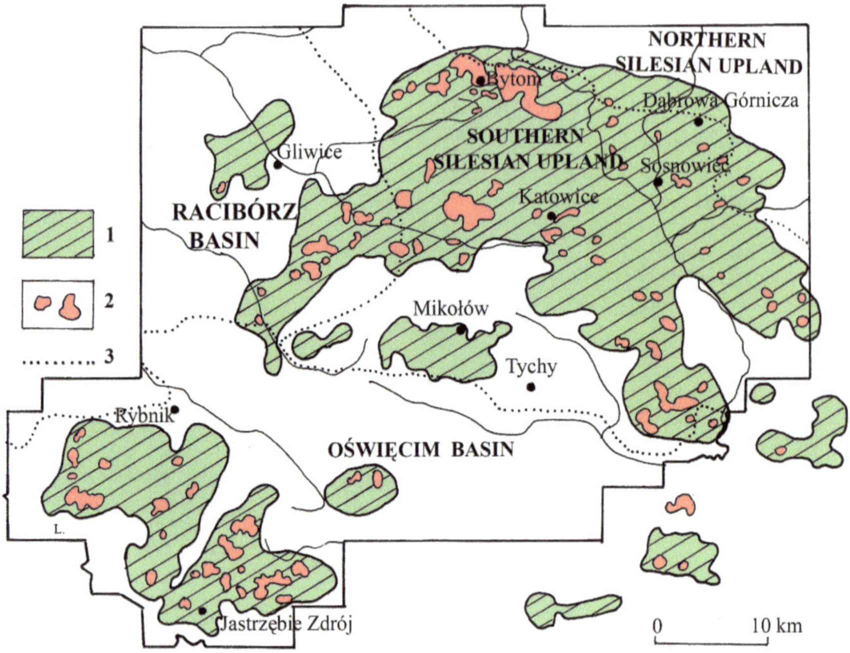

Fig. 3.11 Range of subsidence in the Upper Silesian Coal Basin in the early 1990s (based on Perski 2000). (**1**) Range of subsidence, (**2**) areas lowering at a speed greaterthan 10 mm/month in October 1992, (**3**) boundaries of geomorphological mesoregions

for 1883 showed the relief from the post-mining period because 1.2 % of coal was extracted in the USCB at the time, The DEM model for 1993 showed the relief after the period of the most intense mining; until then, 83.4 % of the total coal mining by 2009 had been extracted.

The volume of subsidence troughs together with excavations that were formed in the study area (2,838 km^2) in the period from 1883 to 1993 amounted to 4,157 million m^3, including 3,284.2 million m^3 in the area of coal mines (1,604 km^2) and 3,339.8 million m^3, when taking into account historical mines and expanding the area of mines to 1,660 km^2 (Table 3.4). In this area, 97.4 % of the volume of anthropogenic concave forms are subsidence troughs created as a consequence of coal, zinc, and lead mining; 2.3 % are sandpits and only 0.3 % are quarries and clay pits. The distribution of subsidence troughs was analysed against the background of the Carboniferous, Triassic, and Miocene geological zones.

In the **Carboniferous zone** (outcrops of carbon with or without the Quaternary deposits cover), mining activities were, for a long time, conducted at a minor depth in the oldest mines in the USCB: Murcki, Wawel, Bolesław Śmiały, Paryż, Polska, Niwka-Modrzejów, Katowice, Wieczorek, and Mysłowice. In this zone, the greatest volume of subsidence was reported, amounting to nearly 1.24 billion m^3. Continuous deformations covered almost 75 % of the area, causing its lowering by

Table 3.4 Volume of depressions caused by subsidence and surface mining and average surface lowering to 1993 according to geological zones (from Dulias 2013)

Characteristic	Carboniferous zone	Triassic zone	Miocene zone	Mining area[a]
Area (km^2)	516.4	284.2	859.4	1660
Percentage of mining area (%)	31.1	17.1	51.8	100.0
Total volume of depression (mln m^3) caused by mining of coal and Zn–Pb ores, stowing sands, solid rocks, and cohesive rocks	1310.1	934.2	1095.5	3339.8
	1235.8	925.4	1093.2	3254.4
	71.8	2.8	1.0	75.6
	–	5.8	–	5.8
	2.5	0.2	1.3	4.0
Percentage of total volume of depression	39.2	28.9	32.8	100.0
Subsidence area (km^2)	385.7	213.7	526.0	1125.4
Average surface lowering (m)	3.4	4.4	2.1	3.0

[a]Mining area together with the historical exploitation fields

an average of 3.2 m. The lowering is predominantly up to 1 m (23 % of the mining impact area), while subsidence exceeding 15 m occurred in the area of 4 km^2, representing only 1 % of the mining impact area.

To the greatest extent, continuous deformations affected the Ruda Hills, which constitutes a compact group of heights bound by the river valleys of the Bytomka, the Bielszowice Stream, and the Rawa. In their western part, the Saddle Beds were exploited, with particularly thick coal seams. In a continuous mining period exceeding 260 years, approximately 1 billion tonnes of coal and waste rock were extracted. The loss of 260 million m^3 of rock material caused the lowering of more than 84 % of the Hills −5.7 m on average. More than 42 % of the area of mining impact lowered by 5–10 m. The maximum surface lowering amounted to over 25 m. Mine subsidence is clearly visible in the relief of the Chorzów Hills, which lowered by an average of 3.6 m. The largest subsidence of about 10–15 m is found in the north-western part of the area of the Śląsk-Matylda mine and partly in the Wawel and Szombierki mines. In the area of the Kochłowice Hills, the greatest subsidence zone (7–14 m) includes mainly two denudation monadnocks located within the impact of the Wujek, Kleofas, and Śląsk mines. The western part of the Hills and their slope toward the Podstokowa Zone decreased by 5–10 m. The whole area of this geomorphological unit lowered by an average of 3.2 m. In the Murcki Plateau, large subsidence affected its central part, mainly the Bolina catchment, but the biggest subsidence occurred in the southern part of the Wesoła mine, where the top of one of the hills in the area of the Książęcy fault subsided by 20–30 m. The Murcki Plateau subsided by an average of 4.1 m.

In the two Basins located in the Carboniferous zone—Mysłowice and Biskupi Bór—an average surface subsidence amounted to 3.9 and 2.6 m, respectively. In this area, however, not all subsidence is associated with continuous deformation resulting from coal mining. Here, intensive mining of stowing sands was carried out and rock material loss related to this type of mining is responsible for an average

Table 3.5 Subsidence in the mining area of the Upper Silesian Coal Basin according to geological zones (from Dulias 2013)

Subsidence (m)	Upper Silesian Coal Basin			Carboniferous zone (km^2)	Triassic zone (km^2)	Miocene zone (km^2)
	(km^2)	(%)	Cumulative percentages			
0–1	345.2	30.7	30.7	88.5	49.5	207.2
1–2	211.9	18.8	49.5	67.3	32.7	111.9
2–3	152.8	13.6	63.1	56.2	25.6	71.0
3–4	105.9	9.4	72.5	43.8	20.1	42.0
4–5	94.5	8.4	80.9	40.6	19.2	34.7
5.0–7.5	111.7	9.9	90.8	47.8	28.2	35.7
7.5–10.0	53.2	4.7	95.5	25.1	14.0	14.1
10.0–12.5	23.8	2.1	97.6	9.6	8.2	6.0
12.5–15.0	11.0	1.0	98.6	2.8	6.0	2.2
15–20	9.2	0.8	99.4	1.8	6.3	1.1
20–25	3.6	0.3	99.7	1.0	2.5	0.1
25–30	1.8	0.2	99.9	0.7	1.1	–
>30	0.8	0.1	100.0	0.5	0.3	–
Total	1125.4	100.0	100.0	385.7	213.7	526.0

subsidence of 0.5 m in the Mysłowice Basin and 1.4 m in the Biskupi Bór Basin (which is more than due to coal mining −1.2 m).

In the **Triassic zone** (an area where Carboniferous carbon is covered by Triassic rocks with zinc and lead ore deposits), mining activities were conducted in numerous mines: Siemianowice, Pstrowski, Barbara-Chorzów, Rozbark, Centrum, Szombierki, and Powstańcow Śląskich. At the beginning of the twentieth century, coal mining expanded to the north, where new mines were established: Miechowice, Bobrek, Andaluzja, Jowisz, and the youngest Julian (1955). The Triassic zone also includes larger or smaller parts of mines of the Dąbrowa Basin. Three quarters of the zone are within the range of mining impact—the volume of subsidence is over 925 million m^3, so the area subsided by an average of 4.3 m and almost 25 km^2 subsided by more than 10 m (Table 3.5). Large mining subsidence is caused, to a certain extent, by the imposition of coal mining impact and impact from zinc and lead ore mining.

The largest subsidence was found in the north-eastern part of the Miechowice Upland near Bytom-Miechowice, which amounted to about 35 m (Solarski and Pradela 2010) (Figs. 3.12 and 3.13). The volume of subsidence in this Upland is almost 400 million m^3. The area of mining impact subsided by an average of 5 m, with about one-fifth of the Upland by more than 10 m. An equally large subsidence was observed in the Siemianowice Upland, with an average value of 5.1 m (subsidence volume of almost 350 million m^3). More than a quarter of the mining impact area subsided by 5–10 m, with widespread and deep (25–34 m) subsidence troughs occurring primarily in the northern part of the Upland (Table 3.6).

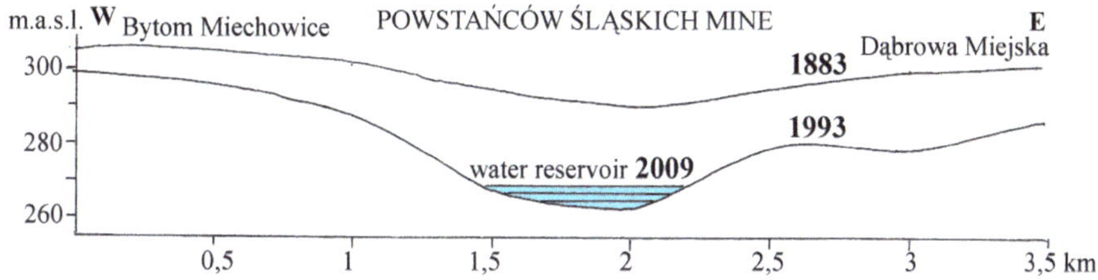

Fig. 3.12 Morphological profile through the deepest subsidence trough in the area of the Upper Silesian Coal Basin, Bytom Miechowice (see Figs. 3.13 and 3.14—profile drawn roughly diagonally across the photo) (from Dulias 2013)

Fig. 3.13 The deepest subsidence trough in the area of the Upper Silesian Coal Basin, Bytom Miechowice (Dulias 2009)

In the area of the remaining geomorphological units of the Triassic zones, mining subsidence is relatively lower. The Czeladź Upland subsided by an average of 2.4 m (of which 2.2 m was due to coal mining) and the Wojkowice Hummock by 2.7 m, mainly in the Brynica valley. A distinct characteristic of the placement of the largest continuous deformation in the Triassic zone is their same shape arrangement with respect to the Saddle Beds (Fig. 3.14).

In the **Miocene zone** (an area where carboniferous carbon is underneath impermeable, hundreds-of-meter-thick Miocene sediments), mining activities have been conducted in 25 coal mines; as many as 13 mines are new, formed after the 1960s. In the area within the range of mining impact (526 km^2), the volume of subsidence is nearly 1,100 million m^3; the surface lowered by an average of 2.1 m (Tables 3.4 and 3.5).

Table 3.6 Subsidence in the selected geomorphological units (from Dulias 2013)

Geomorphological units	Percentage share of subsidence							
	0–1 m	1–2 m	2–3 m	3–4 m	4–5 m	5–10 m	10–15 m	>15 m
Carboniferous zone								
Ruda Hills	10.1	8.1	8.4	8.8	10.6	42.1	10.1	1.8
Chorzów Hills	18.3	17.3	15.0	12.4	11.3	21.0	4.6	0.1
Kochłowice Hills	19.4	17.7	14.9	13.0	12.5	21.0	1.4	0.1
Murcki Plateau	29.6	25.0	18.7	10.9	8.3	6.5	0.4	0.2
Mikołów Hummock	44.6	24.6	13.6	7.2	4.9	4.8	0.3	–
Mysłowice Basin	16.8	13.1	14.6	13.1	15.6	22.7	2.8	1.3
Triassic zone								
Miechowice Upland	35.7	11.7	5.7	4.5	5.5	18.1	10.5	8.3
Siemianowice Upland	12.5	11.9	12.3	12.2	12.4	26.9	7.2	4.6
Czeladź Upland	29.2	21.9	16.9	10.6	8.0	12.2	1.0	0.2
Bobrowniki Hills	42.5	23.0	13.7	8.4	6.2	6.2	–	–
Wojkowice Hummock	21.1	22.5	18.5	11.9	10.6	14.2	1.2	–
Miocene zone								
Kłodnica Graben	52.0	21.0	12.2	6.2	5.2	3.3	0.1	–
Southern Podstokowa Zone	45.8	18.9	13.5	7.7	6.5	7.1	0.5	–
Rachowice High Plain	64.0	12.8	6.8	4.2	3.6	6.9	1.6	0.1
Rybnik Plateau	35.1	19.1	14.1	9.1	8.8	11.5	1.8	0.5

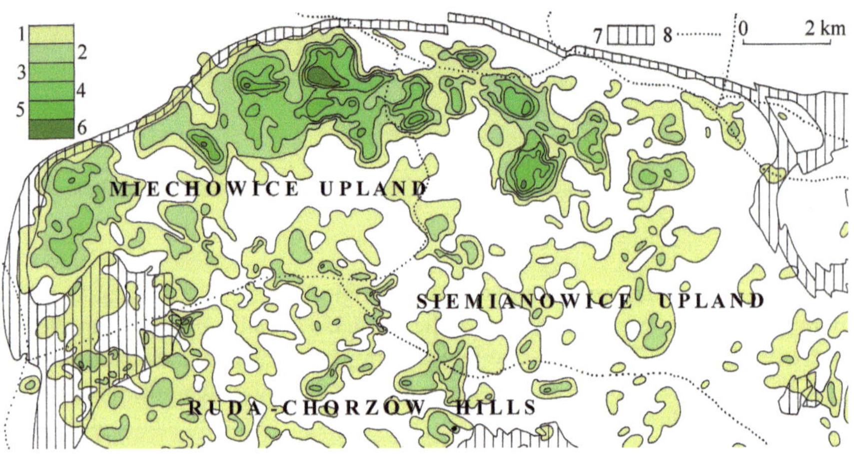

Fig. 3.14 The largest subsidence troughs in the area of the Upper Silesian Coal Basin (from Dulias 2013, generalized, without depressions of up to 5 m). Subsidence. (**1**) 5–10 m, (**2**) 10–15 m, (**3**) 15–20 m, (**4**) 20–25 m, (**5**) 25–30 m, (**6**) more than 30. Other explanations: (**7**) Saddle Beds (Upper Carboniferous), (**8**) boundaries of geomorphological units

Half of the mines in the Miocene zone are located in the Rybnik Plateau. These mining activities contributed to the lowering of its surface by an average of 2.4 m. Subsidence troughs occur here in varied morphological situations—on hilltops, slopes, and valleys. The largest, with depths of 20–25 m, were found within the mines of Chwałowice and Jankowice. Large subsidence occurred in the Rachowice Plateau and the Southern Podstokowa Zone. On average, they lowered by 1.3 and 1.7 m, respectively; however, many of subsidence troughs have a depth of over a dozen meters (Table 3.6). In the area of the Northern Podstokowa Zone near Zabrze-Makoszowy, the maximum subsidence until 1993 amounted to 11–12 m. In the eastern part of the Miocene zone, the average surface subsidence within the range of mining activities of the Ziemowit and Piast mines is 5–7 m.

In the period of 1994–2009 (not covered by morphometric analysis), the emergence of continuous deformations was largely reduced due to the elimination of numerous mines and reduced output. The surface subsided significantly on active mining fields. Substantial subsidence occurred in Ruda Śląska. However, an example particularly fraught with consequences was the district of Bytom-Karb, where serious mining damage forced the eviction of hundreds of people. Exploitation of protective pillars under city districts and industrial facilities conducted since the mid-twentieth century, regarded as a great achievement of Polish mining, did not survive the test of time in all places. Large surface subsidence was found in the vicinity of Sosnowiec-Maczki, in the eastern part of the Kazimierz-Juliusz mine. Since 1996, roof-collapsing exploitation of the 510 seam was carried out here under the surface, where several railway lines had been developed above. On the basis of geodetic measurements, it was found that within 1 year (2001–2002), the studied section of a railway line subsided by 2.55 m (Kowalski and Tracz 2003). In the following years, the increase in subsidence forced a continuous renovation and upgrading of railway tracks. Significant surface subsidence occurred in the western part of the USCB, in the area of the Szczygłowice (1961) and Budryk (1994) mines established in the post-war period.

3.4 Sinkholes

In Poland, discontinuous deformations (superficial and linear) occur in areas of shallow exploitation of hard coal and lignite, ores, raw materials, and rock salt. The boundary between shallow and deep exploitations is generally determined to be at a depth of 80–100 m. Discontinuous deformations, due to their generally small size, are relatively rarely presented on topographic maps. In the area of the Upper Silesian Coal Basin, the emergence of discontinuous deformations is related both to the exploitation of coal and zinc and lead ores. Shallow coal exploitation was carried out in the nineteenth century, particularly in the Main Saddle area, and was continued in some areas until the mid-twentieth century. Discontinuous deformations occur mainly in the north-central part of the USCB, in the strip from Dąbrowa Górnicza to Zabrze (Goszcz et al. 1991). Smaller areas covered by discontinuous deformations occur at the Rybnik Plateau and the Mikołów Hummock (Janusz et al. 1982). On the other hand, shallow mining of zinc and lead ores resulted largely from the occurrence of deposits at shallow depths: discontinuous surface deformations occur in areas of Bytom, Bobrowniki and Chrzanów-Jaworzno. Ore exploitation in coherent and durable grey dolomite (sulphide deposits) favoured the preservation of voids in the bedrock for a long period of time; on the other hand, deposit exploitation, mainly calamine, within red dolomites with a low hardness resulted in a rapid collapse and deformation of the surface (Chudek et al. 1998).

At the turn of the nineteenth and twentieth centuries, sinkholes were often recorded in various written sources. In 1888, Ryszkiewicz wrote that "the road leading through Dąbrowa to Będzin collapsed in several places. On its sides, houses were half-ruined, deserted due to the slow collapse of the exploited mine under them (…). Such sites, called sinkholes, are fenced around" (Zieliński 1984). This and other descriptions of the landscape of the coal basin of the period, through mainly shallow operations, show that the discontinuous deformations constitute clearly visible elements of the landscape. Indirectly, this indicates the intensity of their formation during this period. At the end of the twentieth century, the area of discontinuous deformations occupied 350 km^2 (Liszkowski 1991).

Sinkholes have been inventoried and analysed several times in scientific studies (Chudek and Olaszowski 1976; Palki 1978; Chudek and Arkuszewski 1980; Goszcz et al. 1991). At the end of the twentieth century, in the USCB, about 1,000 discontinuous deformations were registered (Goszcz 1996), but their actual number was and is much greater. There is a lack of information about the number of deformations created in the initial phase of mining, and also most of the mines did not begin to keep inventory until the postwar period.

A statistical analysis was performed on a set of 417 discontinuous deformations that emerged mainly in the years 1960–1980 in areas of shallow coal mining in the mining areas of 27 mines (Goszcz et al. 1991). The analysis leads to the following conclusions: most deformations (about 92 %) were sinkholes and most of them were established during the Cracow Sandstone Series, in which a significant percentage were comprised of permeable sandstone of relatively low mechanical strength. A number of sinkholes, however, emerged in the areas of Saddle Beds of the Upper Silesian Sandstone Series, in which sandstone was characterized, in turn, with a high mechanical strength (Kotyrba 2005). As many as 84 % of the sinkholes resulted from operations conducted at a depth up to 60 m, and none were formed in connection with operations at depths greater than 100 m (in the analysed set, whereas in other works such cases have been recorded). More than one-third of all sinkholes occurred during the period up to 25 years from the completion of mining operations, and 16 % in the interval of 80–100 years (twice more than in the range of 60–80 years) (Goszcz et al. 1991). In a later work, Goszcz (1996) reported about sinkholes formed after 100 years. In general, the vast majority of sinkholes occurred in areas where the Carboniferous roof is covered by Quaternary deposits, and a smaller portion occurred in areas where it is covered by Triassic rocks. The set described above is not complete: it does not include data from the Rybnik district and areas of imposition of coal mining and bullion, and sinkholes that occurred on wasteland or forest land and have not been reported anywhere (Kotyrba 2005).

Sinkholes occur mainly in the Southern Silesian Upland, particularly in the Bytom Plateau and in Mysłowice Basin, and on the Katowice Plateau and the Mikołów Hummock. In the Northern Silesian Upland, sinkholes are found in the Biskupi Bór and Dąbrowa Basins and certain parts of the Middle Triassic Cuesta. In the Racibórz-Oświęcim Valley, sinkholes mainly occur on the Rybnik Plateau; in the Cracow Upland (within the USCB), they occur in the Myślachowice Hills. The characteristics of the sinkhole regions in particular geological zones are presented here.

In the **Carboniferous zone**, discontinuous deformations occur in many areas due to the fact that easy access to coal enabled shallow exploitation almost in the entire area for a long time. In the Chorzów Hills, the problem of sinkholes appeared in the nineteenth century. As the Dziennik Poznański reported in 1879 for the area of Królewska Huta, "a shaft collapsed together with a field through which the railway to Bytom runs, so that transportation was interrupted for a few days. Later on the ground swallowed two houses" (Zieliński 1984). One hundred years later, the problem of sinkholes along said track was still valid: in the years 1966–1976, 11 sinkholes were found there, and a strong reactivation of workings came as a result of operating at a depth of 120 m (Chudek and Arkuszewski 1980). Due to a high concentration of forms along the tracks and the train station, it was concluded that an additional factor activating the rock mass were vibrations of the heavy rolling stock. The total number of registered sinkholes amounted to 27, with an average diameter of 4 m; shallow excavations of former exploitation within the Carboniferous uplift, the so-called Chorzów dome, sunk. A sinkhole area (approximately 1 km^2), considered by the authors as typical, occurred in the southern part of the Barbara-Chorzów mine. Nearby, also within the Chorzów dome but in the eastern part of the Pokój mine (the former President mine), dozens of discontinuous deformations were formed. Sudoł and Zych (2006) describe sinkholes formed in the years 1948–2005 on a flat hilltop (2 km^2). The Carboniferous bedrock here is broken up with numerous faults, which, together with fissures and cracks formed in the overburden, make up a hydraulic string that allows rainwater and surface water to flow freely into the rock mass. Half of at least 46 deformations were created between 1962–1974, mostly in spring periods in connection with the thaw, groundwater melting, and rainfall. Some forms repeatedly reactivated in the same locations—even seven times over several years for one of the sinkholes. The area is still threatened by the creation of sinkholes with diameters that could theoretically exceed 6 m.

In the area of the Ruda Hills, sinkholes only began to be inventoried since 1975. In their western part (the Zabrze-Bielszowice mine) in the years 1975–1996, a total of 11 sinkholes formed in old or badly liquidated shafts, including mine declines (Zych et al. 2000). In Ruda Śląska, on the other hand, 32 forms were inventoried in an area of shallow exploitation of over 10 km^2, mainly within the Wawel mine (Foryś and Surowiec 1985).

On the Dańdówka Plateau, sinkholes formed in an area of about 3 km^2 (Atlas geologiczno-inżynierski 2005; Dulias 2008). They were shaped like cones, but also more irregularly because collapses occurred frequently over the initial decline sections as a result of negligence in their casing. This area is still threatened by the formation of discontinuous deformations. In the northern part of the Porąbka-Klimontów mine, in the area of 1.3 km^2, a total of 248 underground workings have been found with connections to the surface, with an additional 42 in the southern parts (Duży et al. 2000). In the case when there is claystone of poor strength over the workings, which is susceptible to water activity, the voids mostly close up; however, when there is coherent, thick-bedded sandstone, the voids may persist for a long time and constitute a potential sinkhole threat.

The Murcki Plateau's area of the former shallow sinkhole-threatening exploitation is about 13 km^2 (Staniek and Kupka 2007). The sinkholes are present within a few mines: Murcki, Wesoła, Staszic, Wieczorek, and Mysłowice, and mostly in forested areas. The 23 voids at depths of 3–34 m in the districts of Katowice—Nikiszowiec, Giszowiec, and Murcki—might be hazardous to people (Atlas geologiczno-inżynierski 2005).

A lot of sinkholes were formed in the Mysłowice Basin, where shallow deposits were exploited within the Niwka-Modrzejów, Jan Kanty, and Sosnowiec mines and partly in the Jaworzno mine (Fig. 3.15). In the northern sandy part of the Basin, 9 large sinkholes were formed in the early twentieth century—the largest of which had a diameter of nearly 100 m (Mapa Zagłębia Dąbrowskiego 1929). In the 1930s, there were many illegal (poverty) coal pits in the Basin, which also collapsed. Most of the dozens of sinkholes were formed in the years 1964–1971, despite the fact that they resulted from operations carried out in the years 1860–1920—more than 100 years earlier in some cases (Chudek and Arkuszewski 1980; Foryś and Surowiec 1983). An exceptionally intensified emergence of discontinuous deformations characterized the central and south-eastern part of the Mysłowice Basin. The sinkhole area in the district of Sosnowiec-Niwka was recognized by Chudek and Arkuszewski (1980) as typical in terms of the shape of the 23 recorded forms. At its extension to the south and east from Jęzor to the districts of Jaworzno (Dąbrowa Narodowa and Niedzieliska), there was even a greater accumulation of sinkholes—at least 42 with 3–6 m and larger diameters. In this area, belonging to the Jan Kanty mine, as many as a dozen mine declines were formed in the 1950s and 1960s, which produced even up to 40 % of the total output of the mine (Lewandowska 1969). The sudden inflows of water led to frequent collapses of workings and the tunnelling phenomena in the sandy overburden, giving rise to sinkholes on the surface. Wilk (2003) reported that in 1984, irruption of water took place in the outcrop of the Przemsza fault, which led to the formation of a very large sinkhole with almost vertical walls, a depth of 41 m and a diameter of 25 m. The form was eliminated but reactivated three more times; a new crater was formed in its vicinity, with a diameter of 13 m and a depth of 7 m. One of the largest sinkholes formed in 1964 in the southern part of the Mysłowice Basin in the district of Jaworzno–Podłęże, as a result of sandwater flow to the workings through a fault fissure; it was exceptionally large in diameter at about 100 m (Fig. 3.16). The

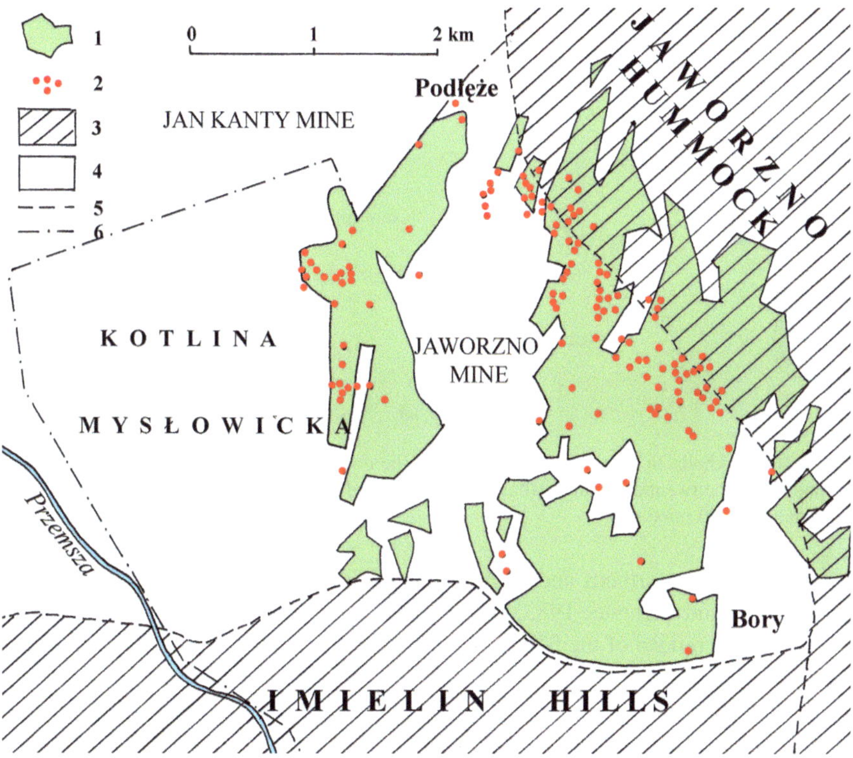

Fig. 3.15 Areas of shallow coal mining and sinkholes in the Jaworzno Mine in Mysłowice Basin (based on Jarczyk 2007). (**1**) Areas of shallow coal mining, (**2**) sinkholes, (**3**) Carboniferous bedrock beneath Triassic overburden, (**4**) Carboniferous bedrock beneath Quaternary deposits, (**5**) boundaries of geomorphological units, (**6**) boundaries of coal mines

sinkhole took the shape of a cone, with a capacity of 75,000 m^3 (Wilk 2003). In this area, a number of typical sinkholes shaped like cones were inventoried within the Carboniferous sandy overburden.

In the Mikołów Hummock, 4 sinkhole areas were distinguished within the activities of the Bolesław Śmiały mine, covering about 6 km^2 (Chudek and Arkuszewski 1980). In two areas, sinkholes appeared as a result of reactivation of workings from the turn of the nineteenth and twentieth centuries. The remaining two were formed a year or two after the exploitation carried out in the late 1950s and 1960s.

In the Biskupi Bór Basin, discontinuous deformations occurred in areas of shallow exploitation carried out in the Kazimierz-Juliusz and Siersza mines. The volume of underground workings within this second mine remaining after shallow operation in the years 1879–1939 was calculated at 1.3 million m^3; 90 % of them, in the 1980s, were completely inflowed by water (Foryś and Surowiec 1983). Many sinkholes formed in the districts of Sosnowiec (Ostrowy Górnicze and Kolonia

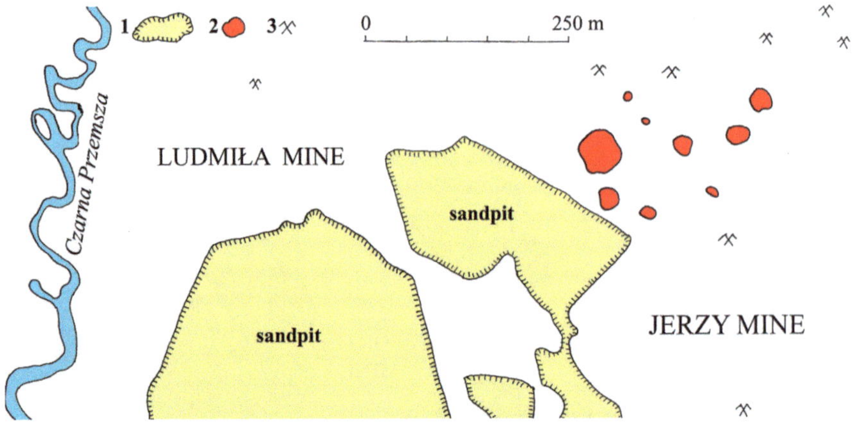

Fig. 3.16 Sinkholes in Sosnowiec-Modrzejów in the northern part of the Mysłowice Basin at the beginning of the twentieth century (based on Mapa Zagłębia Dąbrowskiego 1929). (**1**) Sandpits, (**2**) sinkholes, (**3**) mine shafts

Bory)—9 in the northern and 14 in the southern part (Chudek and Arkuszewski 1980; Foryś and Surowiec 1983). The latter formed in the early twentieth century as a result of operations of the 510 coal seam at a thickness of almost 7 m in the years 1890–1901; these sinkholes have been preserved until today.

In the southern part of the Biskupi Bór Basin, in the area of the Siersza mine, sinkholes already existed in the early twentieth century in the vicinity of the Kozi Bród valley. In 1922, following a violent downpour, the river flooded over the Izabella mine and reached the underground workings through sinkholes and sand sediments, causing the death of 28 miners (Pietraszek 1961; Wilk 2003). The event took place during the operation of a 5-m-thick seam at a depth of 60 m, with the roof-collapsing method under sands with a thickness of 40 m. At present, in the vicinity of the Kozi Bród valley, there is a vast field (0.5 km^2) with more than 30 cones of an average depth of 8–10 m and a diameter of 15 m. These forms were created as a result of the rainwashing of sandy surface sediments to shallowly located workings (Nieć et al. 2001; Głogowska 2007).

In the Dąbrowa Basin, shallow operation was carried out in the Paryż and Grodziec mines. Large sinkholes formed in 1969 as a result of water and silt flow from the settling tank to the Paryż mine, washing the stowage. More than a dozen forms were created within clay overburden resulting from the collapse of workings from the first half of the twentieth century, as well as in the areas of the illegal (poverty) coal pits. Within the limits of the Grodziec mine, located at Grodziec Hights, discontinuous deformations were registered in the area of the former Maria mine that extracts coal under Saint Dorota Mount. Twelve sinkholes formed in the years 1900–1936 (Chudek and Arkuszewski 1980), but the particular intensity of their formation falls within a period of intense exploitation in the years 1967–1975. In extreme cases, the formation of sinkholes with a depth of several meters with a large diameter was observed, including the uncovering of old underground workings. During the flooding of the Maria mine, which began in 1983, such phenomena were not observed; it was not until the flooding was complete that the emergence of new forms has been observed every few years (Bukowski and Augustyniak 2005).

An interesting example of discontinuous deformations comes from the area of the Myślachowice Hills on the Ojców Plateau (Ostrowski and Mularz 1987). Due to operations carried out in the Siersza mine at a depth of 130 m in a previously intact rock mass, two large 120-m-long sinkhole trenches appeared. At first, parallel fissures formed within a few meters of each other, and the area between them collapsed to a depth of several meters. The uneven bottom had alternating deep wells and bridges. These forms were created on the hilltop of a longitudinal elevation (420 m above sea level), built from Myślachowice conglomerates. Under these forms and on the Carboniferous top, an arkose series composed mainly of sandstone is found, with inserts of clay forms. As a result of long-term tunnelling and the leaching of poorly cohesive overburden rock, it was strongly caverned and rock mass deformation caused by roof-collapsing operations contributed to a sudden tightening of caverns and the creation of sinkhole forms.

In the **Triassic zone**, many discontinuous deformations were created on the Siemianowice Upland. In its northern part, where the impact of zinc and lead ore mining overlaps with the impact of coal mining, their genesis may be directly related to the exploitation of ores or the exploitation of coal seams located deeper, which causes the reactivation of post-ore workings. Until the mid-1970s within the Orzeł Biały Mining and Metallurgy Plant, 177 sinkhole forms were inventoried in an area of 5.8 km^2, with an average diameter of 4–5 m (Chudek and Arkuszewski 1980). According to Pilecki (2009), until 1995, deformities associated with ore mining prevailed here; only later were they connected with the mining of coal. In the area of the Rozbark coal mine in the period of 1959–1994, only 7 sinkholes were directly caused by ore exploitation, whereas 46 were connected with the reactivation of old workings as a result of coal mining (Chudek et al. 1998). In the years 1974–1990, due to the overlap of the Orzeł Biały Mining and Metallurgy Plant with the mining work of the Andaluzja mine, 91 deformations emerged; 75 of them appeared on the surface in a period of 2 years following the completion of ore mining (Fig. 3.17). The size of the forms was varied: the maximum recorded diameter was 28 m and the maximum depth was 7 m (Dulias 2013). In three cases, sinkholes reached the underground workings. Some of them still exist today, despite being 20 years since their formation; in wet seasons, they periodically fill with water. In the eastern part of the Siemianowice Upland within the Saturn mine, many sinkholes appeared at the turn of the nineteenth and twentieth centuries as a consequence of coal mining with the checkerboard extraction method (pillar retaining exploitation in wide galleries) (Ciepiela 2003).

In the Czeladź Upland, discontinuous surface deformations were created mainly in its southern part within the Czeladź-Miłowice mine. Numerous sinkholes appeared as early as in the nineteenth century as a result of 60-m-deep roof-collapsing extraction (Ciepiela 2003). In the 1970s, a total of 68 sinkholes were inventoried in the area, of which 30 originated in the valley of Brynica, with

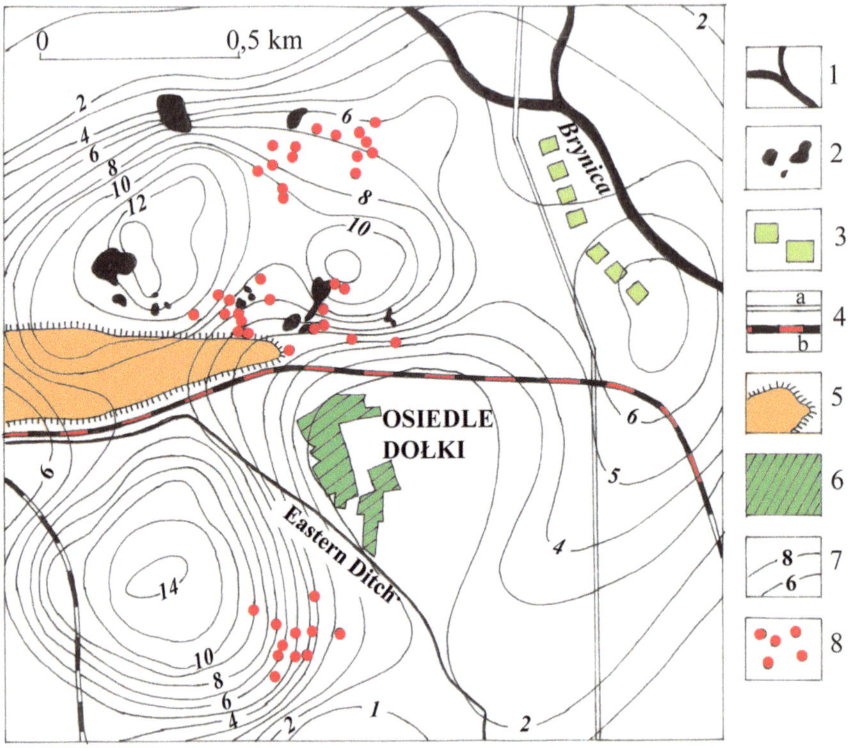

Fig. 3.17 Sinkholes in subsidence trough in the area of overlapping influences of the Andaluzja and Orzeł Biały mines in northern part of the Siemianowice Upland (from Dulias 2013). (**1**) Rivers, (**2**) water reservoirs, (**3**) settling tanks, (**4**) a main roads, b railways, (**5**) mining spoil tip, (**6**) residential areas, (**7**) isolines of surface subsidence in meters, (**8**) sinkholes

significant participation of water infiltrating the pits; their average diameter was 8.3 m (Chudek and Arkuszewski 1980). The sinkhole area covered about 2 km^2. As many as 44 sinkholes were created over several years (1957–1965) as a result of operations conducted in mine declines (Foryś and Surowiec 1983).

In the Tarnowskie Góry Plateau, sinkholes frequently appear in the proximity of shafts. They have a depth of up to 5 m, a width of up to 25 m, and occur in high density in the areas of Górniki, Repty, or Segiet (Lamparska-Wieland 2003). These forms are associated with zinc and lead ore mining.

In the **Miocene zone** (an area where Carboniferous is present under a cover of Miocene rocks), sinkholes are not found; however, they occur in a small area where the Carboniferous bedrock is under a cover of Quaternary deposits. In the mining area, mostly within the Rybnik Plateau, 469 discontinuous deformations were recorded, including 242 sinkholes (Palki 1978). Almost 97 % of them occur in the eastern part of the Rydułtowy mine, while 8 forms are scattered in the areas of 5 mines. In 1956, within the limits of the Jankowice mine, a large crater was formed due to the breaking of water and sand into the workings. The crater had a depth of

17 m and a diameter of 50 m (Wilk 2003). The largest concentration of sinkholes occurs in a small area where Quaternary deposits, with a thickness of 10–40 m in the form of alternating layers of sand and gravel and clay, are located on the Carboniferous rocks (Palki 1982). The most common forms of sinkholes are cones (190) and cylindrical-cones (45), which account for 97 % of all sinkholes (Janusz and Palki 1980). Almost half of the deformations had a diameter of 5–10 m; slightly more than 42 % had a diameter of 10–20 m. The maximum diameter of a form was 67 m. The depth of sinkholes ranged from 0.8 to 23.5 m, while up to 71 % of forms had a depth of 3 m (Palki 1982). Some forms that have been preserved have water reservoirs, such as in the area of the Anna mine located in the Syrynka valley (5 m deep) or in the Valley of the Chwałowicki Stream (Jankowski 1986). The largest dimensions were characterized by sinkholes created due to damages to hydrated Carboniferous overburden forms resulting from mining works. In these cases, the workings were located at a depth of even 110 m. For all sinkholes, the geological and mining conditions of their creation were analysed (Palki 1981). It was acknowledged that in 94 % of cases, the reason for their formation was the exploitation with the roof-collapse method of shallowly located seams; in 3 % of cases, the reactivation of old shallow workings was due to various reasons.

In the Upper Silesian Coal Basin, voids located in the rock mass may persist indefinitely; only in appropriate conditions does the void begin to wander up (Goszcz 1996). Many original voids have not yet collapsed, nor have they been stowed during a mine closure. In probabilistic terms, the risk of sinkholes in the USCB area is small and comparable to other coal basins (Kotyrba 2005; Strzałkowski et al. 2006). Nevertheless, almost every year there are isolated cases of sudden sinkholes above former underground workings.

The calculation of the amount of material displaced from the surface through sinkhole forms to the underground workings is difficult. By assuming that 1000 cone-shaped sinkholes with an average depth of 3 m and a diameter of 6 m were formed in the area of the USCB, it would give a material volume of about 28,000 m^3. With the inclusion of approximately 100 more sinkholes of a larger size at a depth of 15 m and diameter of 30 m, the volume reaches over 350,000 m^3. Taking into account the opinions that the actual number of deformities in the USCB is much higher than 1000 and some cones have a capacity of 75,000 m^3 (Wilk 2003), it seems that the estimated volume of the displaced material at 1 million m^3 is not overstated.

3.5 Fissures and Thresholds

Until the 1990s, discontinuous linear deformations statistically accounted for only about 8 % of the total discontinuous deformations found in the area of former shallow coal mining (Goszcz et al. 1991). On the Rybnik Plateau in the 1960–1980s, linear deformations accounted for 48 % of all registered forms (Palki 1978, 1981). Currently, linear deformations occur more frequently than sinkholes, which is encouraged by the nature of the operations—roof-collapsing on several seams up to one border designated by a protective pillar, a fault, or a mining area (Strzałkowski et al. 2006).

In the Rybnik Plateau, almost 98 % of linear deformations were surface thresholds and fissures, usually occurring simultaneously. Fissure width ranged from several millimetres to about 0.6 m and the thresholds reached 0.8 m in height (less frequently, 1.2 m). Cracks were found occasionally, possibly due to their fast seizure caused by their insignificant size (Palki 1981). Linear deformations occurred when the operation was performed at a wide range of depths, up to 600 m, under a thick overburden of both the Carboniferous as well as Tertiary and Quaternary rocks (Janusz et al. 1982). Two examples of deformations created between 1978 and 1979 come from areas of the Pniówek and Zofiówka mines. The first of these had a series of fissures and arch-shaped thresholds that followed the outline contour of the mining fields. The main fissure was about 600 m long and 0.55 m wide, and the accompanying threshold was 0.2 m high. Parallel to this deformation at distances of 10–20 m, there were signs of cracks, fissures, and thresholds of slightly smaller sizes. The apparent depth of the cracks reached up to 0.9 m. In the second case (the Zofiówka mine), linear deformations formed in an area 40–100 m wide. The length of fissures with the associated surface thresholds (up to 0.5 m in height) here ranged between 100 and 500 m. The emergence of deformations was associated with the speed of the exploitation face formation and the presence of desiccated clay and silt near the surface (Janusz et al. 1982).

In the middle part of the Rybnik Plateau (around Marklowice) in the years 1986–2003, many discontinuities were inventoried within the noncohesive and locally cohesive soils with water-bearing levels at shallow depths (Kowalski 2005a). Altogether, 53 steps (single or composed of several steps) had heights ranging between 0.05–0.7 m and lengths from several to 250 m. The width of the fissures ranged between 0.02 and 0.2 m.

In 2002–2003, Kowalski (2005b) studied three regions with identified discontinuities. In each, the overburden is composed of Quaternary deposits with a thickness of 5–30 m located directly on the Carboniferous rock with faults in the roof. The fissures have a length from 25 to 100 m, a width of 0.5 m, and depth of up to 1.5 m. Six flexures were also found at a length of 230 m, with a height of 0.1–0.2 m.

In the area of the Knurów mine (located in the town of Gierałtowice within the Rachowice Upland), numerous surface thresholds formed in the last 30 years due to the exploitation carried out at depths up to 720 m (Kruczkowski 1999; Strzałkowski et al. 2006). Three zones of deformations have been described here, the largest of which (approximately 200 m wide) is in the centre of the village (Kruczkowski 1999). According to the inventory of 1997, a total of 51 surface thresholds formed here with heights up to 0.5 m, which were revealed in stages during 1990–1995. After 2000, further linear deformations appeared, with heights of up to 0.5–0.6 m (Strzałkowski et al. 2006). Deformations occurred near the overlapping edges of several seams; their parallel alignment refers to the parallel arrangement of tectonic faults. One of the causes of the deformations is believed to be the presence of nondurable Tertiary clay in the overburden (Kruczkowski 1999).

References

Atlas geologiczno-inżynierski aglomeracji katowickiej 1:10 000. Katowice, Warszawa, Wrocław 2005

Białek J, Mielimąka R (1999) Skutki szybkiej eksploatacji prowadzonej w jednej z kopalń ROW dla zabudowy jednorodzinnej. Zesz Nauk Politech Śląskiej, Górnictwo 239:65–75

Bukowski P, Augustyniak I (2005) Analiza zjawisk związanych z zaprzestaniem odwadniania wyrobisk górniczych na przykładzie byłej kopalni Maria. Bezp Pracy Ochr Środ Górn 1 (125):13–20

Chudek M, Arkuszewski J (1980) Identyfikacja deformacji zapadliskowych w obszarach dawnej i płytkiej eksploatacji górniczej na terenie Górnośląskiego Okręgu Przemysłowego. Projekty – Problemy Budownictwa Węglowego 4:9–16

Chudek M, Chudek MD, Moj H (1998) Zagrożenie deformacjami nieciągłymi w rejonach pokrywających się obszarów górniczych ZGH "Orzeł Biały" i KWK "Rozbark". Mat VIII Międzynar Symp Geotechnika'98, Gliwice-Ustroń, pp 107–120

Chudek M, Olaszowski W (1976) Określenie rodzaju i wielkości deformacji nieciągłych powierzchni. Ochr Teren Górn, p 38

Ciepiela B (2003) Najstarsze i ostatnie kopalnie węgla w Zagłębiu Dąbrowskim, czyli końcowa synteza zagłębiowskiego górnictwa. Stow. Autorów Polskich, Sosnowiec

Dulias R (2008) Płytka eksploatacja węgla kamiennego na Płaskowzgórzach Dańdówki. Acta Geogr Silesiana 3:13–17

Dulias R (2010) Landscape planning in areas of sand extraction in the, Poland. Landscape Urban Plan 95(3):91–104

Dulias R (2011) Impact of mining subsidence on the relief of the, Poland. Zt Geomorph 55, Suppl 1, Stuttgart, Januar, pp 25–36

Dulias R (2013) Denudacja antropogeniczna na obszarach górniczych na przykładzie Górnośląskiego Zagłębia Węglowego. Wyd Uniw Śląskiego, Katowice

Duży S, Kleta H, Plewa F (2000) Zagrożenie powierzchni ze strony starych zrobów w obszarze likwidowanej kopalni. Zesz Nauk Politech Śląskiej, Górnictwo 246:111–117

Foryś T, Surowiec Z (1983) Metody geofizyczne i ich przydatność w ocenie zagrożenia powierzchni przez pustki poeksploatacyjne na terenie Sosnowca. Mat Konf - Ochrona środowiska na terenie miasta Sosnowca. PTPNoZ Oddz Górnośląski, Warszawa-Sosnowiec, pp 95–111

Foryś T, Surowiec Z (1985) Możliwość wykorzystania metod geofizycznych do oceny zagrożenia powierzchni przez pustki i inne nieciągłości górotworu na przykładach prac wykonanych w Rudzie Śląskiej. In: Konstantynowicz E (ed) Ochrona środowiska naturalnego na terenie aglomeracji miejskiej Rudy Śląskiej. PTPNoZ, Oddz Górnośląski, Sosnowiec, pp 93–104

Gawor Ł, Szmatłoch A (2010) Koncepcja waloryzacji zwałowisk po górnictwie węgla kamiennego na przykładzie Górnośląskiego Zagłębia Węglowego. Acta Geogr Silesiana 7:23–29

Głogowska M (2007) Walory edukacyjne odsłonięć geologicznych i obiektów górniczych w okolicach Trzebini. AGH, Kraków (dissertation)

Goszcz A (1996) Powstawanie zapadlisk i innych deformacji nieciągłych powierzchni na obszarach płytkiej eksploatacji górniczej. Szkoła Ekspl Podziem, Szczyrk, pp 119–137

Goszcz A, Surowiec Z, Kotyrba A, Foryś T (1991) Analiza metod i możliwości oceny oraz sposoby zwalczania zagrożenia powierzchni ze strony płytko zalegających pustek. Prace GIG 763, Katowice

Górecki J, Szwed E (2005) Pozostałości dawnego górnictwa kruszcowego na ziemi krzeszowickiej. Prace Nauk Inst Górn Politech Wrocł 111, Konferencje 43

Greszta J (1957) Z frontu walki z nieużytkami poprzemysłowymi. Chroń Przyr Ojczystą 5:3–12

Hejmanowski R, Malinowska A (2009) Evaluation of reliability of subsidence prediction based on spatial statistical analysis. Int J Rock Mech Min Sci 46:432–438

Helios-Rybicka E, Rybicki S (2002) Environmental impacts of coal mining in Poland. In Puura E, Marmo L, D'Alessandro M (eds) Mine and quarry waste—the burden from de past. Inst Environ Sustain, Orta Italy, pp 35–37

Hornig A (1968) Wpływ działalności gospodarczej człowieka na środowisko geograficzne GOP. Czas Geogr 39(1):13–26

Jania J (1983) Antropogeniczne zmiany rzeźby terenu wschodniej części Wyżyny Śląskiej. Dokum teledet. Prace Nauk Uniw Śląskiego 575:69–91

Jankowski AT (1986) Antropogeniczne zmiany stosunków wodnych na obszarze uprzemysłowionym i zurbanizowanym (na przykładzie ROW). Uniw. Śląski, Katowice

Janusz W, Palki J (1980) Propozycje klasyfikacji nieciągłych deformacji powierzchni terenu w obszarach eksploatacji górniczej. Ochr Teren Górn 52

Janusz W, Palki J, Zygmunt J, Węgrzyk A (1982) Występowanie deformacji nieciągłych przy dużych głębokościach eksploatacji w Rybnickim Okręgu Węglowym. Mat. Konf. Zagadnienie ochrony powierzchni w ROW, Jastrzębie Zdrój

Jarczyk MJ (2007) Przywracanie wartości użytkowych terenu metodami wiertniczymi w otoczeniu likwidowanych kopalń. AGH, Kraków (doctoral dissertation)

Karaś-Brzozowska M (1960) Charakterystyka geomorfologiczna Górnośląskiego Okręgu Przemysłowego. Kom. d/s GOP PAN, Biul. 37

Kotyrba A (2005) Zagrożenie i ryzyko zapadliskowe terenów Górnośląskiego Zagłębia Węglowego. Wiad Górn 7–8:348–358

Kowalski A (1996) Wpływ eksploatacji na deformacje powierzchni na przykładzie eksploatacji KWK Staszic. Mat Konf Szk Ekspl Podziem, Katowice, pp 139–152

Kowalski A (2005a) Rozpoznanie i możliwości prognozowania liniowych deformacji nieciągłych powierzchni. In: Kwiatek J (ed) Problemy eksploatacji górniczej pod terenami zagospodarowanymi. VIII Dni Miern Górn Ochr Teren Górn, Ustroń, pp 278–291

Kowalski A (2005b) O liniowych nieciągłościach powierzchni. Bezp Pracy Ochr Środ Górn 12:25–33

Kowalski A, Tracz P (2003) Kształtowanie się deformacji powierzchni spowodowanej eksploatacją drugiej warstwy pokładu węgla kamiennego w kopalni "Kazimierz-Juliusz" – wyniki eksperymentu. Geodezja 9, 2/1:357–367

Kruczkowski M (1999) Deformacje nieciągłe typu liniowego na terenie miejscowości Gierałtowice jako przejaw intensywnej eksploatacji górniczej w niekorzystnych warunkach geologicznych. Bezp Pracy Ochr Środ Górn 7(8):41–45

Kupka R, Szczypek T, Wach J (2005) Morphological effect of 200-years long hard coal exploitation in Katowice. In Szabó J, Morkūnaitė R (eds) Landscapes – nature and man. Univ. of Debrecen. Lithuanian Inst. of Geol Geogr, Debrecen-Vilnius, pp 95–100

Lamparska-Wieland M (2003) Atlas zmian wybranych elementów krajobrazu terenów górniczych i pogórniczych Płaskowyżu Tarnowickiego. Prace WNoZ UŚ 27, Warszawa

Lamparska-Wieland M, Waga JM (2002) Significance of slag dumping areas in the Upper Silesia and West Małopolska . Anthr aspects of land transform 2:32–35

Lewandowska J (1969) Ziemia chrzanowska i Jaworzno Monografia. Wyd. Literackie, Kraków

Liszkowski J (1991) Engineering and environmental impacts caused by land subsidence due to subsurface extraction of solid raw materials from Poland. IAHS Publ 200:369–377

Machowski R (2010) Przemiany geosystemów zbiorników wodnych powstałych w nieckach osiadania na Wyżynie Katowickiej. Wyd. Uniw. Śląskiego, Katowice

Machowski R, Rzętała MA, Rzętała M (2006) Procesy i formy brzegowe w obrębie jeziora poeksploatacyjnego w początkowym okresie funkcjonowania na przykładzie zbiornika Kuźnica Warężyńska. Kształt środ geogr ochr przyr obsz uprzem zurban 37:29–36

Mapa geomorfologiczna Górnośląskiego Okręgu Przemysłowego 1:50 000. Klimaszewski M (ed.) Kom dsGOP PAN, Warszawa 1959

Mapa przeobrażeń powierzchni ziemi woj. katowickiego 1:50,000. WOS UW Katowice 1982

Mapa Zagłębia Dąbrowskiego 1:10000. Woj. Inst Geogr 1929

Mielimąka R (2006) Pomierzone i prognozowane krzywizny terenu górniczego na przykładzie obserwacji geodezyjnych z KWK "Budryk". Górn Geol 1(4):81–92

Nieć M, Kawulak M, Salamon E (2001) Mapa geologiczno-gospodarczo-sozologiczna 1:25 000 dla miasta i gminy Trzebinia. PAN IGSMiG, Kraków

Ostrowski J, Mularz S (1987) Znaczenie rozpoznania warunków geologicznych dla trafnego prognozowania deformacji nieciągłych. Ochr Teren Górn 81/3, 82/4:24–29

Palki J (1978) Powstawanie deformacji nieciągłych powierzchni terenu przy prowadzeniu robót górniczych pod zawodnionymi utworami w stropie karbonu. Ochr Teren Górn 45

Palki J (1981) Zakres, formy i przyczyny występowania deformacji nieciągłych w warunkach Rybnickiego Okręgu Węglowego. Ochr Teren Górn 56

Palki J (1982) Zagadnienie zagrożenia powierzchni zapadliskami w Rybnickim Okręgu Węglowym. Mat Konf VIII Tydzień Techniki ROW, Jastrzębie Zdrój, pp 102–125

Pełka J, Pociecha J (1991) Zmiany terytorialne w rozmieszczeniu zwałowisk górniczych na obszarze Rybnickiego Okręgu Węglowego w latach 1861–1986. Kształt środ geogr ochr przyr obsz uprzem zurban 2:5–10

Pełka-Gościniak J (2006) Restoring nature in mining areas of the (Poland). Earth Surf Proc Land 31(13):1685–1691

Perski Z (2000) Zastosowanie satelitarnej interferometrii radarowej do określania dynamiki i zasięgu górniczych deformacji terenu na przykładzie wybranych obszarów Górnośląskiego Zagłębia Węglowego. Prace WNoZ UŚ 8, Sosnowiec, pp 9–39

Perski Z, Jura D (1999) ERS SAR interferometry for land subsidence detection on coal mining areas. Earth Obs Q 63:25–29

Pietraszek E (1961) Ośrodek górniczy Siersza 1804–1861–1961. Wyd. Artyst Graf, Kraków

Pietrzyk-Sokulska E (1995) Wpływ podziemnej eksploatacji i przeróbki węgla kamiennego na środowisko przyrodnicze w Polsce. Stud Rozp Monogr, 39. Wyd. CPPGSMIE PAN, Kraków, pp 5–77

Pilecki Z (2009) Methodology for A1 motorway basement treatment effectiveness improvement by means of geophysical methods in the areas of metal ores shallow minning threatened with the sinkhole occurrence in the Upper Silesia. Gosp Sur Mineral 25(3):319–331

Popiołek E, Ostrowski J (1981) Próba ustalenia głównych przyczyn rozbieżności prognozowanych i obserwowanych poeksploatacyjnych wskaźników deformacji. Ochr Teren Górn 58

Rechowicz H (1974) Kopalnia Generał Zawadzki. Dzieje zakładu i załogi. Śląski Inst Nauk, Katowice

Rzętała M (2008) Funkcjonowanie zbiorników oraz przebieg procesów limnicznych w warunkach zróżnicowanej antropopresji na przykładzie regionu górnośląskiego. Wyd. Uniw. Śląskiego, Katowice

Rzętała MA (2003) Procesy brzegowe i osady denne wybranych zbiorników wodnych w warunkach zróżnicowanej antropopresji (na przykładzie Wyżyny Śląskiej i jej obrzeży). Wyd. Uniw. Śląskiego Katowice

Skarżyńska KM, Burda H, Klepacz J (1988) Obiekty budownictwa wodnego z nieprzepuszczalnych odpadów kopalni węgla kamiennego. Gosp Wodna 3

Skinderowicz B (1982) Wskaźniki deformacji dynamicznych niecek osiadania. Mat. konf. Zagadnienia ochrony powierzchni w Rybnickim Okręgu Weglowym, Jastrzębie Zdrój, pp 21–40

Solarski M, Pradela A (2010) Przemiany wybrany form rzeźby Wyżyny Miechowickiej w latach 1883-1994. Z badań nad wpływem antropopresji na środowisko 11:78–92

Staniek F, Kupka R (2007) Płytka eksploatacja węgla kamiennego w graniach miasta Katowice. Informacja ogólna. Kształt środ geogr ochr przyr na obsz uprzem zurban 38:42–45

Starościak E (2006) 280 lat historii Dąbrowy Górniczej. W 90-lecie uzyskania praw miejskich. Muzeum Miejskie "Sztygarka", 10. Dąbrowa Górnicza

Strzałkowski P, Piwowarczyk J, Łapajski K (2006) Występowanie deformacji nieciągłych liniowych w świetle analiz warunków geologiczno-górniczych. Przegl Górn 5:1–5

Sudoł A, Zych J (2006) Analiza deformacji nieciągłych na terenie byłej kopalni "Prezydent". Górn Geol 1(4):101–116

Szczypek T, Wach J (1991a) Human impact and intensity of in the Silesian-Cracow Upland (Southern Poland). Z Geomorph NE, Suppl.-Bd. 90, Berlin-Stuttgart, pp 171–177

Szczypek T, Wach J (1991b). Rozwój współczesnej wydmy w warunkach silnej antropopresji. Wyd. Uniw. Śląskiego, Katowice

Szczypek T, Wach J (1993) Antropogenicznie wymuszone procesy i formy eoliczne na Wyżynie Śląskiej. SGP, Poznań

Szczypek T, Wach J (1999) Human impact and development of a modern scarp dune. In Schirmer W (ed) Dunes and fossil soils. GeoArchaeoRhein, Münster 3:177–186

Wach J. 1991. Wpływ antropopresji na kształtowanie się rzeźby terenu. Mat. Symp. polsko-czeskiego – Człowiek i jego środowisko w górnośląsko-ostrawskim regionie przemysłowym. UŚ WNoZ Sosnowiec, pp 115–119

Wilk Z (ed) (2003) Hydrogeologia polskich złóż kopalin i problemy wodne górnictwa. 1, Uczel Wyd. Nauk-Dydakt, Kraków

Wojciechowski T (2007) Osiadanie powierzchni terenu pod wpływem eksploatacji węgla kamiennego na przykładzie rejonu miasta Knurowa. Przegl Geol 55(7):589–594

Zieliński A (1984) Górny Śląsk i Zagłębie w dawnych opisach - wiek XIX. Wyd. Śląsk Katowice

Zuchiewicz W (2000) Współczesne ruchy tektoniczne. Mapa 1: 5 000 000. In Atlas Polski. Encyklopedia geograficzna świata, Wyd. Opres, Kraków, p 29

Zych J, Baranowski Z, Dźwigoł H (2000) Problemy ochrony powierzchni po zlikwidowaniu kopalni na przykładzie byłej kopalni "Zabrze". Zesz Nauk Politech Śląs. Górnictwo 246:407–417

Żmuda S (1973) Antropogeniczne przeobrażenia środowiska przyrodniczego konurbacji górnośląskiej. PWN Warszawa-Kraków

Chapter 4
Changes in Morphometric Parameters of Terrain Caused by Mining

The contemporary relief of the Upper Silesian Coal Basin is similar to the relief from the pre-mining period. However, even a rough comparison of contour maps from the late nineteenth century with contemporary ones highlights inferior differences, such as the appearance of anthropogenic landforms, changes in the geometry of river beds, changes in altitudes, and others. Cartographic materials dating back to the nineteenth and twentieth centuries may therefore be used in studies of relief transformations occurring in the past 100–150 years because they allow for an objective evaluation and comparison of areas. Morphometric analyses based on archival and contemporary topographic maps are particularly useful in the study of areas subject to strong anthropogenic pressure. Such studies have been conducted by many authors for different parts of the Upper Silesian Coal Basin and are mostly based on morphological relief profiles, valleys' longitudinal profiles, and hypsographic curves prepared with the use of topographic maps issued in different years (e.g. Jankowski 1986; Wach 1987; Szczypek and Wach 1996; Madowicz 2001; Dulias 2003, 2005, 2008, 2011, 2013; Aleshina et al. 2008; Kupka et al. 2008; Solarski and Pradela 2010). In this chapter, attention is focused on the impact of mining on changes in the three qualities of the USCB relief: absolute altitudes, relative altitudes and slope inclinations. The results of digital elevation models for 1883 and 1993 were analysed in relation to 3 geological zones and 25 geomorphological units. The latter were divided into three groups: those with a dominant share of plains (with a surface inclination of less than 1°), those with a dominant share of slightly and gently inclined slopes (1°–5°), and those with a substantial share of slopes, at least moderately inclined (above 5°). Many geomorphological units located within the Upper Silesian Coal Basin are characterized by an almost flat nature of relief; they primarily include larger erosion-denudation and tectonic depressions, such as the Basins of Biskupi Bór, Mysłowice, Mleczna, the Grabens of Kłodnica and Chrzanów, as well as the even High Plains of Rachowice, Czechowice and Wilcza and the elongated N-S Podstokowa Zone on the border of the Silesian Upland and the Racibórz Basin. In all of these units, plains occupy from 50 % to even 88 % of their area.

R. Dulias, *The Impact of Mining on the Landscape*,
Environmental Science and Engineering, DOI 10.1007/978-3-319-29541-1_4

4.1 Changes in Altitude

In the USCB, changes in absolute altitudes are mainly related to mining subsidence and cover vast areas. They were examined in two aspects: in terms of changes of areas in particular altitude intervals, as well as changes in an average altitude of the terrain. In the Carboniferous area, the biggest changes of absolute altitudes concern the heights of 240–250 m above sea level. The area with such heights increased by over 12 km^2, whereas the area with an altitude of 310–320 m above sea level decreased by 6 km^2 (Table 4.1). In the Miocene zone, the biggest changes of absolute altitudes took place in relation to altitudes of 250–260 m above sea level, whose area decreased by almost 21 km^2, whereas the area with altitudes of 220–240 m above sea level increased by 25 km^2. Changes in the Triassic zone varied, with a clear increase of more than 20 km^2 for areas located at altitudes of 260–280 m above sea level; on the other hand, the areas of higher altitudes, ranging from 290 to 310 m above sea level, decreased by nearly 23 km^2.

In the entire mining area, the area with high altitudes of above 280 m decreased by over 42 km^2, with an increase of areas located below 250 m by nearly 34 km^2. In addition, areas located at 250–260 m above sea level decreased by more than 18 km^2, with an increase of areas with altitudes 260–280 m above sea level by almost 26 km^2. In individual research units, absolute altitudes changed differently—the biggest changes included the lowest and/or highest hypsometric levels or were related mainly to the medium-altitude zone. In the areas of the largest mining

Table 4.1 Changes in altitude in the mining area according to the geological zones in the period of 1883–1993 (after Dulias 2013)

Altitude (m above sea level)	Area (km^2)					
	Carboniferous zone		Triassic zone		Miocene zone	
	1883	1993	1883	1993	1883	1993
Below 220	–	–	0.2	0.2	4.5	5.5
220–230	–	0.035	0.3	0.4	16.7	28.3
230–240	0.8	4.1	0.4	0.8	71.2	84.6
240–250	11.3	23.4	6.2	6.7	158.5	149.9
250–260	51.6	52.4	11.2	12.9	174.4	153.7
260–270	84.8	83.1	25.9	33.8	144.2	150.0
270–280	90.5	92.9	39.6	52.0	141.0	140.0
280–290	87.2	85.2	69.6	73.4	103.8	101.6
290–300	67.1	63.8	68.8	56.0	31.4	31.3
300–310	51.2	48.2	37.1	27.0	9.9	9.6
310–320	36.3	30.3	13.3	11.4	2.6	2.5
320–330	20.8	17.8	6.5	5.3	0.9	1.1
330–340	8.7	9.1	3.3	2.9	0.3	0.6
Above 340	6.1	5.8	1.6	1.2	–	0.1

subsidence, the changes in the absolute altitudes very clearly reflect the hypsometric maps and morphological profiles for 1883 and 1993.

An average altitude above sea level of three geomorphological units—the Mleczna Basin, the Kłodnica Graben, and the Bobrowniki Hills—increased by 0.3–0.4 m due to the formation of large spoil tips and, in the case of the Bobrowniki Hills, a settling tank and embankments along the Brynica River. However, the altitude did not change in the Czechowice Upland at all. The remaining 21 units experienced a decrease of their average altitude by 0.2–4.5 m (Table 4.2). The largest reduction in the average altitude was recorded in the Ruda Hills (4.5 m), in the Siemianowice Upland (4.1 m), and the Miechowice Upland (3.3 m). It is characteristic that in the period 1883–1993, the smallest changes in the average altitude of the terrain were recorded for geomorphological units that were predominantly plains and units with a high proportion of slopes with an inclination of more than 5°. The biggest changes (threefold) took place in units with a dominant share of slopes with an inclination of 1°–5°.

The difference between average altitudes above sea level for 1883 and 1993 is reflected in the size of an average decrease/increase of the height during this period. These values were calculated for the whole geomorphological units; therefore, they are not consistent with the values for surface lowering given in Sect. 3.3, which were calculated for the area of mining influence in these units.

Table 4.2 Changes in average altitude of selected geomorphological units in the years 1883–1993 (after Dulias 2013)

Geomorphological units	Average altitude (m)		Increase/decrease of average altitude in the years 1883–1993 (m)
	1883	1993	
Geomorphological units with a dominant share of plains (more than 50 % of area)			
Biskupi Bór Basin	269.5	268.1	−1.4
Mysłowice Basin	257.8	255.6	−2.2
Mleczna Basin	252.9	253.3	+0.4
Kłodnica Graben	267.0	267.4	+0.4
Chrzanów Graben	247.7	245.4	−2.3
Czechowice High Plain	235.0	235.0	–
Rachowice High Plain	241.0	240.3	−0.7
Wilcza High Plain	242.9	242.7	−0.2
Northern Podstokowa Zone	239.8	238.9	−0.9
Southern Podstokowa Zone	252.5	252.0	−0.5
Average	250.6	249.9	−0.7

(continued)

Table 4.2 (continued)

Geomorphological units	Average altitude (m)		Increase/decrease of average altitude in the years 1883–1993 (m)
	1883	1993	
Geomorphological units with a dominant share of slopes inclined by 1°–5° (more than 50 % of area)			
Miechowice Upland	283.5	280.2	−3.3
Siemianowice Upland	280.7	276.6	−4.1
Ruda Hills	280.2	275.7	−4.5
Chorzów Hills	287.2	284.7	−2.5
Czeladź Upland	273.9	272.5	−1.4
Kochłowice Hills	279.8	277.3	−2.5
Murcki Plateau	286.0	283.7	−2.3
Western Mikołów Hummock	284.9	284.5	−0.4
Eastern Mikołów Hummock	298.8	298.3	−0.5
Average	283.9	281.5	−2.4
Geomorphological units with a high share of slopes with an inclination greater than 5° (more than 20 % of area)			
Rogoźnik Hills	313.9	312.2	−1.7
Bobrowniki Hills	288.1	288.4	+0.3
Grodziec Elevations	298.6	297.6	−1.0
Wojkowice Hummock	279.9	279.0	−0.9
Dańdówka Plateau	278.9	278.0	−0.9
Rybnik Plateau	263.2	262.8	−0.4
Average	287.0	286.3	−0.7

4.2 Changes in Relative Heights

From a morphogenetic point of view, relative heights are among the most important features of the relief. The position of the local base-levels of erosion in relation to the hilltops affects the size and the intensity of erosion and denudation processes.

Changes in relative heights in the USCB are associated with mining subsidence, the creation of high spoil tip heaps, and deep excavations. An analysis of these changes in the period 1883–1993 was carried out for 25 geomorphological units located entirely or in a large part within the impact of coal mining. It shows that at the end of the nineteenth century, the maximum denivelation ranged from 38 m in the Chrzanów Graben to almost 117 m in the Grodziec Hills. Over a hundred years later, in 1993, these denivelations ranged from 48.8 m in the Wojkowice Hummock to 212 m in the Rybnik Plateau. Apart from two units that did not record any changes in the maximum denivelations (the Miechowice Upland and the Wilcza High Plain) and two units where the denivelations decreased by 1–1.2 m (the Rachowice High Plain and the western part of the Mikołów Hummock), they increased in all other units, ranging from 2.2 to 95.6 m (Table 4.3 and Fig. 4.1).

Table 4.3 Changes in relative heights in selected geomorphological units in the years 1883–1993 (from Dulias 2013)

Geomorphological units	Maximum relative height (m)			Percentage of area with changes in relative heights		
	1883	1993	Increase/decrease in the years 1883–1993	−10 to 0 m	0.1–5 m	5.1–10 m
Geomorphological units with a dominant share of plains (more than 50 % of area)						
Biskupi Bór Basin	43.0	70.0	+27.0	13.7	38.4	30.2
Mysłowice Basin	70.0	98.0	+28.0	11.8	28.2	23.5
Mleczna Basin	41.1	79.8	+38.7	23.8	55.7	17.9
Kłodnica Graben	65.4	67.6	+2.2	19.4	47.2	24.1
Chrzanów Graben	38.0	83.4	+45.4	15.4	41.6	32.8
Czechowice High Plain	73.8	85.1	+11.3	10.8	65.0	18.5
Rachowice High Plain	82.8	81.6	−1.2	11.9	68.6	14.0
Wilcza High Plain	69.6	69.6	–	4.4	89.2	5.6
Northern Podstokowa Zone	47.6	51.7	+4.1	10.4	61.9	13.4
Southern Podstokowa Zone	63.8	132.9	+69.1	10.1	54.1	24.7
Average				13.2	55.0	20.5
Geomorphological units with a dominant share of slopes inclined by 1°–5° (more than 50 % of area)						
Miechowice Upland	103.8	103.7	−0.1	25.9	21.3	37.0
Siemianowice Upland	56.9	63.2	+6.3	17.7	27.8	32.9
Ruda Hills	81.0	93.5	+12.5	22.2	33.3	22.2
Chorzów Hills	71.0	79.5	+8.5	32.7	41.4	15.3
Czeladź Upland	50.0	66.8	+16.8	15.7	39.4	21.7
Kochłowice Hills	101.4	111.3	+9.9	29.6	49.0	16.3
Murcki Plateau	102.3	132.6	+30.3	32.8	29.6	23.0
Western Mikołów Hummock	82.3	81.3	−1.0	18.8	38.8	37.2
Eastern Mikołów Hummock	110.0	139.0	+29.0	26.4	39.0	22.0
Average				24.6	35.5	25.3

(continued)

Table 4.3 (continued)

Geomorphological units	Maximum relative height (m)			Percentage of area with changes in relative heights		
	1883	1993	Increase/decrease in the years 1883–1993	−10 to 0 m	0.1–5 m	5.1–10 m
Geomorphological units with a high share of slopes with an inclination greater than 5° (more than 20 % of area)						
Rogoźnik Hills	101.1	105.0	+3.9	60.1	29.8	3.0
Bobrowniki Hills	60.0	68.3	+8.3	30.3	45.5	15.2
Grodziec Elevations	116.9	121.3	+4.4	64.0	33.0	–
Wojkowice Hummock	39.2	48.8	+9.6	11.1	69.4	11.1
Dańdówka Plateau	70.0	73.6	+3.6	24.1	49.1	16.4
Rybnik Plateau	116.4	212.0	+95.6	24.7	49.8	18.5
Average				35.7	46.1	10.7

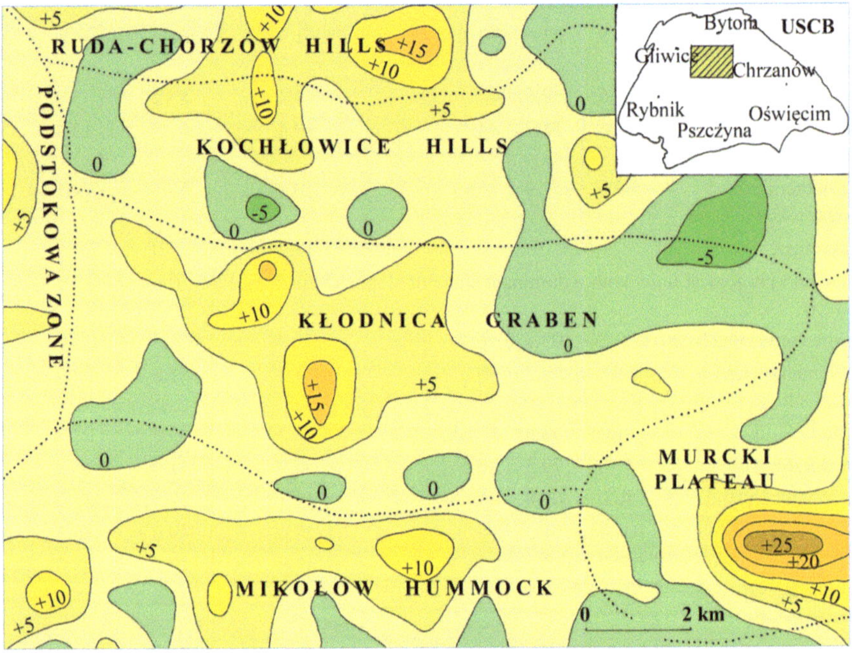

Fig. 4.1 Changes in relative heights in the central part of the Upper Silesian Coal Basin for the period of 1883–1993 (values of isolines in meters) (from Dulias 2013)

A large increase in maximum denivelations occurred mainly in the units dominated by plains (e.g. the Chrzanów Graben + 45.4 m, the Mleczna Basin + 38.7 m), whereas it was insignificant for areas with diverse relief (e.g. the Rogoźnik Hills + 3.9 m, the Grodziec Hills + 4.4 m); an exception is the Rybnik Plateau, where the maximum denivelations increased by almost 100 m.

In the period of 1883–1993, an increase in relative heights took place in as much as 77.3 % of the analysed geomorphological units. For almost 46 % of their area, relative heights increased maximally up to 5 m (mainly on plains); for almost 20 %, the increase ranged from 5.1 to 10 m; and for more than 11 % of the area, it was more than 10 m. In the remaining area, relative heights decreased by up to 10 m; these changes occurred primarily in geomorphological units with a high proportion of slopes inclined more than 5°. In general, due to changes in relative heights, the relief of lowlands became varied, and vice versa: the areas where the relief was diversified were somewhat mitigated. Changes in relative heights in particular geological zones are quite distinctive: the maximum denivelations in the Miocene and Carboniferous zones increased by 42.6 and 56 m, respectively; however, in the Triassic zone, they almost did not change (−0.6 m). An example of an area with significant changes in relative heights during the period 1883–1993 is the central part of the USCB.

4.3 Changes in Slope Inclinations

Slope inclination is one of the most frequently analysed morphometric features of relief. The geomorphological literature does not provide an agreement as to the limits between particular slope inclination categories. In this chapter, the following division of slopes are applied: slightly inclined, 1°–3°; gently inclined, 3°–5°; moderately inclined, 5°–9°; strongly inclined, 9°–20°; and steep, over 20°.

Changes of slope inclinations in the Upper Silesian Coal Basin are associated with three main causes: mining subsidence, the emergence of anthropogenic landforms (e.g., spoil tips, excavations), and changes in topography due to various construction works. In the first case, the slopes change their inclination in large areas, but generally gently (usually about 1°–5°). In other cases, there are slopes with high inclination (a dozen or more degrees), but in small areas. An analysis of changes in the inclination of slopes in the years 1883–1993 shows that they clearly increased in the largest subsidence troughs around the base of spoil tips and the edge of excavations, but also along river valleys (on both sides), which may be explained by natural and anthropogenic factors.

Changes in slope inclination in each of the geological zones have the same characteristics—the arear of plains was reduced with an increase in the size of slopes in each inclination category, and most significantly for slopes in the 1°–3° category (Table 4.4). In the Triassic zone, there was also a quite clear increase of slopes inclined by 3°–5°. In the Carboniferous zone, the area with strongly inclined

Table 4.4 Slope inclinations according to geological zones in the years 1883 and 1993 (from Dulias 2013)

Slope inclination	Area (km²)					
	Carboniferous zone		Triassic zone		Miocene zone	
	1883	1993	1883	1993	1883	1993
Less than 1°	215.1	194.1	114.3	92.0	496.9	429.3
1°–3°	221.9	229.4	114.0	126.2	221.6	268.8
3°–5°	59.0	65.0	34.8	43.3	68.1	77.0
5°–9°	17.6	20.9	17.3	17.7	51.8	57.8
9°–20°	2.7	5.4	3.7	4.2	20.1	24.4
More than 20°	0.1	1.6	0.1	0.8	0.9	2.1

slopes doubled (from 2.7 to 5.4 km^2), partly due to the emergence of numerous spoil tips with slopes of such an inclination.

In the period of 1883–1993 for the majority of the studied geomorphological units, the area of plains decreased by an average of 5.8 %—mostly in the Kłodnica Graben, the Siemianowice Upland, and the Biskupi Bór Basin (Table 4.5). The total area of plains in the research area dropped by more than 113 km^2 (Table 4.6). At the same time, the participation of slopes in all incline categories increased, including mostly slopes with an inclination of 1°–3° (by almost 77 km^2), mainly in the Podstokowa Zone, but also in the Kłodnica Graben and the Biskupi Bór Basin. In addition, the area of gently inclined slopes increased by almost 16 km^2, moderately inclined slopes increased by almost 12 km^2, strongly inclined slopes increased by 7.4 km^2, and steep slopes increased by less than 2 km^2. Slopes with changed inclinations occupy the largest area in the Rybnik Plateau at almost 68 km^2, and the Rachowice High Plain was almost 23 km^2. In percentage terms, the largest changes in slope inclination occurred in the area of the Kłodnica Graben, covering almost 25 % of its surface, as well as in the Siemianowice Upland (24.2 %) and Biskupi Bór Basin (22.4 %).

For geomorphological units that are characterized by almost flat relief, an increase of slopes with an inclination of 1°–3° is not without significance as far as the development of contemporary geomorphological processes is concerned. They currently occupy an area that is about 41 km^2 larger than in the late nineteenth century. Among the geomorphological units with a dominant share of slopes inclined by 1°–5°, two subgroups emerged. The first includes the Miechowice and Siemianowice Uplands and the Ruda and Chorzów Hills. For these areas, an increased inclination of slopes was reported as a result of the increase in slopes with a 1°–3° inclination and larger at the expense of plains. As a result of significant mining subsidence, these units found themselves in a special geomorphological situation, diverse from their previous state. A large part of them is currently land-locked; hence, the increased potential for erosion and denudation is to a large degree directed to the “inside” of the units, rather than to discharge the matter

Table 4.5 Changes in the percentage share of the areas in particular classes of slope inclination in selected geomorphological units in the years 1883–1993 (from Dulias 2013)

Geomorphological unit	Increase/decrease of percentage share of the areas in particular classes of slope inclination					
	<1°	1°–3°	3°–5°	5°–9°	9°–20°	>20°
Geomorphological units with a dominant share of plains (more than 50 % of area)						
Biskupi Bór Basin	−11.2	+7.8	+1.8	+1.0	+0.4	+0.2
Mysłowice Basin	−9.9	+5.9	+1.9	+1.2	+0.7	+0.2
Mleczna Basin	−5.6	+5.4	–	+0.1	+0.1	–
Kłodnica Graben	−12.3	+8.3	+1.9	+1.3	+0.7	+0.1
Chrzanów Graben	−4.7	+3.8	+0.4	+0.2	+0.3	–
Czechowice High Plain	−4.1	+2.3	+0.7	+0.8	+0.3	–
Rachowice High Plain	−4.7	+3.4	+0.7	+0.4	+0.2	–
Wilcza High Plain	−1.4	+0.6	+0.5	+0.2	+0.1	–
Northern Podstokowa Zone	−8.5	+8.5	+0.1	–	–	−0.1
Southern Podstokowa Zone	−8.8	+9.2	−0.8	–	–	+0.4
Geomorphological units with a dominant share of slopes inclined by 1°–5° (more than 50 % of area)						
Miechowice Upland	−4.6	+4.2	+0.7	−0.5	+0.2	–
Siemianowice Upland	−12.1	+4.5	+4.3	+2.0	+0.9	+0.4
Ruda Hills	−1.4	−2.4	+1.9	+0.9	+0.9	+0.1
Chorzów Hills	−0.9	−1.8	+0.8	+0.9	+0.9	+0.1
Czeladź Upland	−4.7	–	+2.5	+1.4	+0.6	+0.2
Kochłowice Hills	−2.4	+1.6	+0.2	+0.4	+0.2	–
Murcki Plateau	−5.3	+5.2	−0.3	+0.3	+0.1	–
Western Mikołów Hummock	−1.7	−2.1	+1.4	+1.7	+0.6	+0.1
Eastern Mikołów Hummock	−2.5	+0.3	+0.8	+0.3	+1.0	+0.1
Geomorphological units with a high share of slopes with an inclination greater than 5° (more than 20 % of area)						
Rogoźnik Hills	+2.1	+1.5	−0.8	−2.3	−1.1	+0.6
Bobrowniki Hills	+6.9	−2.8	−1.8	−3.0	+0.8	−0.1
Grodziec Elevations	+1.9	−1.3	+1.1	−0.5	−1.3	+0.1
Wojkowice Hummock	−10.2	+6.8	+1.5	+0.7	+0.6	+0.6
Dańdówka Plateau	−3.5	+1.5	−2.8	+3.8	+1.4	−0.4
Rybnik Plateau	−7.7	+5.6	+0.9	+0.7	+0.4	+0.1

beyond their borders. The other subgroup includes large geomorphological units—the Murcki Plateau, the Mikołów Hummock, and the Kochłowice Hills. In each of them, there are tectonic faults that are highlighted in the relief in the form of slopes with high inclination; however, in comparison to the large area of these units, the percentage participation of slopes with high inclination is insignificant.

Geomorphological units that are characterized by the most inclined slopes include the Rogoźnik Hills, the Bobrowniki Hills, the Grodziec Hills, the

Table 4.6 Changes in the areas in particular classes of slope inclination in the geomorphological units listed in Table 4.5 for the years 1883–1993 (after Dulias 2013)

Plain/slope	Slope inclination	Increase/decrease of the areas in particular classes of slope inclination (km^2)
Plain	Less than 1°	−113.3
Slightly inclined	1°–3°	+76.8
Gently inclined	3°–5°	+15.6
Moderately inclined	5°–9°	+11.7
Strongly inclined	9°–20°	+7.4
Steep	More than 20°	+1.8

Wojkowice Hummock, and the Dańdówka Plateau—areas located in the zone of the Będzin fault, with a continuous or discontinuous cover of Triassic rocks and lack of or a thin cover of Quaternary deposits. More than 20 % of their area is composed of slopes with an inclination of more than 5°. A particularly strong relationship between the structure of the relief and the slope inclination is noticeable in the case of the Dańdówka Plateau; in this respect, it is a region that stands out in the whole area of research, since more than 46 % of the slopes are moderately or strongly inclined. As a result of mining subsidence, these dependencies were slightly enhanced, both in the case of the Dańdówka Plateau and the Wojkowice Hummock, while they were mitigated in the above-listed hills.

Units with a high proportion of slopes with an inclination greater than 5° (more than 22 % of the area) also include the Rybnik Plateau. The steepness of the slopes in this case is related to their deep fragmentation by a thick network of gullies developed on the loess cover. In the analysed period of 1883–1993, the area of slopes with such an inclination increased by more than 3.5 km^2. A clear reduction occurred in flattened hilltops and valley bottoms by almost 34 km^2 to the benefit of more inclined slopes, which has geomorphological consequences. The slopes of the Rybnik Plateau are now somewhat shorter and steeper, which creates conditions for more intensive erosion processes.

Changes in slope inclination associated with forms of open-pit mining are vast, but mostly include such insignificant surfaces in the scale of the studied regions that they are not always possible to be grasped in a morphometric analysis. Nevertheless, direct field observations indicate that the appearance of anthropogenic landforms in relief with steep slopes (often above 20°) affects the course of geomorphological processes. This applies, above all, to spoil tips that have not been covered with vegetation. At the exit of erosion channels that fragment them, numerous proluvial fans are created; the material that they are composed of is then gradually discharged into the lower parts of the slopes.

References

Aleshina IN, Snytko VA, Szczypek S (2008) Mining induced ground subsidences as the relief-forming factor on the territory of the of the Silesian Upland (Southern Poland). Geogr Nat Resour 29:288–291

Dulias R (2003) Subsidence depressions in Upper Silesian Coal Basin. In: Mentlik P (ed) Geomorf sborník, vol 2, pp 11–16

Dulias R (2005) Krzywe hipsograficzne obszaru osiadań górniczych (na przykładzie okolic Piekar Śląskich). In: Kotarba A, Krzemień K, Święchowicz J (eds) Współczesna ewolucja rzeźby Polski, SGP, IGiGP UJ, IGiPZ PAN, IG AP, Kraków, pp 115–120

Dulias R (2008) Mining subsidence in Oświęcim Basin (Carpathian Foredeep). Geomorphologia Slovaca et Bohemica 8, 2008/2:7–13

Dulias R (2011) Zmiany wysokości względnych na obszarze Katowic w ostatnim stuleciu. Kształt środ geogr ochr przyr obsz uprzem zurban 43:14–19

Dulias R (2013) Denudacja antropogeniczna na obszarach górniczych na przykładzie Górnośląskiego Zagłębia Węglowego. Wyd. Uniw Śląskiego, Katowice

Jankowski AT (1986) Antropogeniczne zmiany stosunków wodnych na obszarze uprzemysłowionym i zurbanizowanym (na przykładzie ROW). Uniw. Śląski, Katowice

Kupka R, Frolik H, Dulias R (2008) Zmiany rzeźby na obszarze górniczym zlikwidowanej kopalni „Katowice-Kleofas". Informacja ogólna. Kształt środ geogr ochr przyr obsz uprzem zurban 39:26–31

Madowicz A (2001) Osiadania terenu na obszarze Jastrzębia Zdroju w latach 1974–1997. Kształt środ geogr ochr przyr obsz uprzem zurban 31:15–21

Solarski M, Pradela A (2010) Przemiany wybrany form rzeźby Wyżyny Miechowickiej w latach 1883–1994. Z badań nad wpływ antropopresji na środowisko 11:78–92

Szczypek T, Wach J (1996) Preobrazovania rel`efa mestnosti v raionakh gornodobyvaiushchei promyshlennosti vsledstvie osedanii grunta (na primere Katovickovo voevodstva). In: Pirozhnik II (ed) Geograficheskie problemy prirodopolzovania v usloviakh antropogennoi deiatelnosti. Belorusskii Gosudarstvennyi univ., Belorusskoe Geograficheskoe obshchestvo, Minsk

Wach J (1987) Zmiany profilu podłużnego Kłodnicy w wyniku osiadań górniczych. Mat. Symp. polsko-czech Problemy geograficzne górnośląsko-ostrawskiego regionu przemysłowego. ODN IKN, UŚ, Katowice-Sosnowiec, pp 126–130

Chapter 5
Changes in the Circulation of Matter in Drainage Basins

Changes in the conditions for the circulation of matter in the USCB are presented separately for open systems (drainage basins) and separately for closed systems (land-locked basins). In open systems, there is a steady supply of energy and matter, internal circulation, and the outflow of energy and matter; in closed systems, there is no drainage or it is very difficult. Mineral and organic matter circulates in the systems as well as substances dissolved in water. The causative agent of the circulation of matter is the circulating water (precipitation, surface flow, midsoil flow, underground runoff), wind, and gravity.

Anthropogenic factors of the circulation of matter in the Upper Silesian Coal Basin result primarily from the long-term effects of mining activities. Nevertheless, there are also other types of anthropopressure present in this area; due to the overlapping of their impact, it is not always possible to clearly identify the source of changes. Mining activities covered regions of different geological structure, relief, water conditions, amount of rainfall, land use, and generally varied potential for erosion and denudation. Changes in the conditions of the circulation of matter in the study area mainly result from the creation of anthropogenic landforms, changes of morphometric features of the relief, restrictions in the infiltration as a result of land development, the emergence of new matter carrier, and changes in the physico-chemical characteristics of surface sediments.

Anthropogenic landforms occur in different locations in relation to natural landforms (valley bottoms, slopes, hilltops, etc.). Their presence influences the course of modern geomorphological processes, mainly slopewash, and fluvial, slope, and aeolian processes (regardless of the fact that these forms are also influenced by these processes). The most significant change in the circulation of matter in the USCB is the formation of numerous land-locked basins, mostly in subsidence troughs and the exclusion of these areas from the fluvial system. The lack of continuity between the slope and fluvial subsystems, as a result of such anthropogenic forms as mine tailings embankments, rail and road embankments, or spoil tips is not without significance. Conditions for the circulation matter in the area of the USCB also changed along with changes of morphometric relief features,

R. Dulias, *The Impact of Mining on the Landscape*,
Environmental Science and Engineering, DOI 10.1007/978-3-319-29541-1_5

as presented in Chap. 4. An increase or a decrease in relative heights, resulting in a change of a "distance" to the local base levels of erosion, may significantly affect the nature and intensity of the flow of energy and matter, especially when combined with changes in the slope inclination and shape. A change in the morphological character of riverbeds and their gradients is of particular importance in this regard (Wach 1987a; Dulias 2008a, 2011).

An important reason for the changing conditions of the circulation of energy and matter in the USCB is the development, paving, and "concreting" of large urban, industrial, and mining areas and thereby a reduction in the infiltration capacity of the soil. New, artificially directed flows of energy and matter are related to the municipal and industrial sewer systems, drainage ditches, canals, and rivers running in closed canals, especially with discharges of large amounts of sewage and mine water into the river system. The most unique feature of the circulation of matter in mining areas is the fact that its important carrier is man, who moves matter against the force of gravity, freely "crossing" natural boundaries of matter circulation systems. Such movement is carried out in different directions—from underground workings to the surface area, from pits to underground workings, from spoil tips into pits, from spoil tips to the surfaces of areas not transformed by mining activities, etc. An exemplary, simplified diagram of the direct movement of rock matter between mines—coal and stowing sand—is shown in Fig. 5.1. Man-manipulated circulation of matter and energy imposes onto the natural circulation that takes place mainly with the participation of water and, to a lesser extent, wind. The spread of polluted air leads to the fall of dust or heavy metals in considerable distances from the sources of their "production."

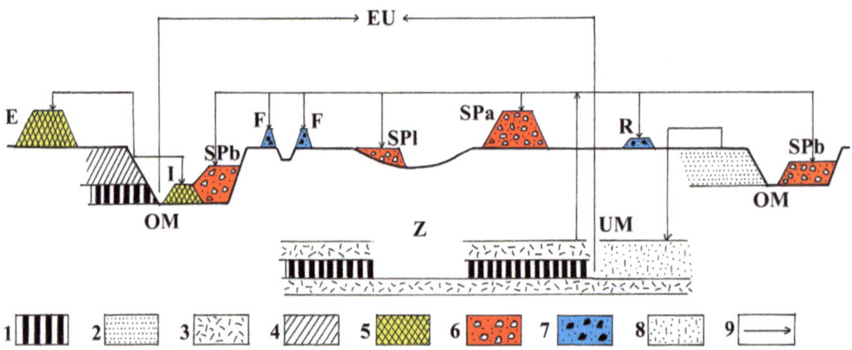

Fig. 5.1 Simplified scheme of the rock material movement between coal and stowing sands mines: (**1**) coal seam, (**2**) stowing sand deposits, (**3**) waste rocks (gangue), (**4**) overburden, (**5**)—external (**E**) and internal (**I**) spoil tips of overburden, (**6**) spoil tips: above ground level (**SPa**), below ground level (**SPb**), levelling heap, for example in a subsidence trough (**SPl**), (**7**) railway (**R**) and flood (**F**) embankments built of mining waste, (**8**) hydraulic stowage, (**9**) direction of rock material movement; (**OM**) open mine, (**UM**) underground mine, (**Z**) fissure and roof-fall zone, (**EU**) economic use

The character of the circulation is also influenced by the physicochemical properties of the moving matter. New lithological deposits appeared on the relief (waste rock, tailings), and the natural landforms were in some areas dried or waterlogged, contaminated with heavy metals, mixed with waste material, deprived of vegetation, etc. The issue is presented in Sect. 5.1.

The amount of material intentionally transported in the USCB can be determined, for example, based on the size of raw material and waste rock extraction, the disposal of tailings and overburden, or the amount of transported suspension in rivers, while data on the amount of material transported due to natural geomorphological processes is modest for this area. In many works devoted to the anthropogenic relief of the Silesian Upland and the neighbouring areas, it is emphasized that mining activities affect the course and intensity of modern geomorphological processes, but only a few quantify the changes (e.g. Jania 1983; Szczypek and Wach 1991a, b; Rzętała M.A. 2003; Kupka et al. 2005; Dulias 2007, 2010). In the future, after the cessation of mining operations and the completion of mine drainage, the circulation of matter from the mining period will be modified due to the filling of the depression cone and, as it is forecast, the hydration or increase in the water inflow in some subsidence troughs and the related changes in local base levels of erosion.

An analysis of changes in the conditions of the fluvial systems functioning under the influence of mining activities in the area of the Upper Silesian Coal Basin was carried out with reference to the fluvial system model by Schumm (1977). It differentiates three zones (subsystems): the production area, which includes the slopes from the watershed to the riverbed; the transfer zone, which is limited by the banks of the riverbed; and the deposition zone, which is in the foreland of river mouths. In the first zone, the river is supplied in energy and matter; in the second, its movement takes place; while in the third, energy reaches a minimum and the matter is accumulated. Each of these zones may be defined as a system or subsystem and is considered as an open system (subsystem). On the basis of archival and contemporary large-scale maps, some parameters of rivers and catchment areas and their changes were presented for the period 1883–1993 (i.e., 111 years), provided that in certain areas the changes occurred in a shorter period of time due to the later initiation of mining operations. The study included 48 rivers of different categories with a total length of over 430 km—29 in the Oder and 19 in the Vistula basin. The basins of these rivers occupy a predominant part of the USCB, where an average of 21 % of their area is located within the mining subsidence greater than 5 m. In relation to certain issues, research results obtained for other rivers and for the period after 1993 were provided, as well as the data from literary sources.

The study area (2,838 km^2) is located almost in equal parts in the basins of the Oder (1,437 km^2) and the Vistula (1,401 km^2). The western part of the area (the Oder basin) belongs mostly to the Miocene zone, while the eastern part (the Vistula basin) belongs mostly to the area of the Carboniferous and Triassic zones. On the scale of the research area, rivers with a length of at least a dozen kilometres were considered large, including the Kłodnica, Bierawka, Brynica, Gostynia, and Mleczna. Medium-sized rivers with lengths of mostly a few kilometers (5–9)—at

Table 5.1 The investigated rivers and streams (from Dulias 2013)

Stream order					Catchment area (km²)
II	III	IV	V	VI	
Vistula Basin					
Przemsza (from the Biała Przemsza mouth to the Imielinka mouth)					215.79
	Brynica (from the Szarlejka mouth)				257.82
		Rawa			83.05
			Leśny Stream		11.61
		Szarlejka			41.75
			Ditch from Radzionków		6.73
		Orzeł Biały Ditch			6.37
		Dąbrówka Wielka Ditch			9.23
		Michałkowice Ditch			23.84
	Bolina				28.14
		South Bolina 1			7.33
		South Bolina 2			6.46
Goławiec Stream					34.42
Gostynia	Mleczna				144.17
		Przyrwa			39.94
			Tributary from Wesoła		5.89
		Murcki Ditch			9.68
Pszczynka	Upper Pszczynka				8.72
	Dębinka				8.19
Oder Basin					881.84
Kłodnica (to the Kozłówka mouth)					523.44
	Bytomka				143.06
		Żernica Stream			65.21
			Tributary in Mikulczyce		2.93
			Tributary from Górniki		6.66
		Miechowice Stream			2.56
	Sośnica Stream				8.74
	Czarniawka				15.29
	Bielszowice Stream				31.94
	Ślepiotka				14.29
	Jasienica	Paniówek Stream			2.61
		Ornontowice Stream	Gierałtowice Stream		4.42
		Tributary from Przyszowice			5.26
	Cienka				11.47

(continued)

Table 5.1 (continued)

Stream order					Catchment area (km²)
II	III	IV	V	VI	
Bierawka (to the Knurówka mouth)					127.10
	Knurówka				18.14
	Szczygłowice Stream				13.28
	Jordanek				10.32
Ruda	Nacyna				68.75
		Pludry			4.43
		Tributary from Michałkowice			21.52
			Tributary from Popielów		9.43
Olza	Szotkówka				162.55
		Stream from Gogołowa			3.11
		Jastrzębianka			10.54
			Tributary from Kąty		2.26
		Lesznica	Marklówka		9.02
			Jedłownik Stream		5.62
				Radlin Stream	5.48

most a dozen or so kilometres (11–13)—included, for example, the Bolina, Szarlejka, Knurówka, and the Szczygłowice Stream. Small watercourses were considered to be the ones with lengths up to 3–4 km, such as the Tributary from Przyszowice, the Tributary in Mikulczyce, the Gogołowa Stream, the Pludry, the Orzeł Biały Ditch, and the Michałkowice Ditch. Many rivers run in a course that is similar to the latitudinal course (e.g. the Bytomka, the Bielszowice Stream, the Czarniawka, the Rawa, the Jordanek) and others have a meridional course (e.g. the Przemsza, the Goławiec Stream, the Szotkówka) (Table 5.1).

Mining activity has had a significant impact on the functioning of fluvial systems. An area of 1,660 km^2 was within its impact, but its effects are transferred outside the area of impact, such as by the river transport of suspension. Changes in the circulation of energy and matter in fluvial systems primarily result from changes in the position of the base level of erosion, the geometry of river channels and their nature (technology development), the inclination of valley bottoms and slopes, the course of watersheds, the resulting changes in catchment areas, and changes in the flow and the load of transported material. These changes are caused by various factors, but especially by mining activities and hydro-technical works. The intensity of the mining impact on the fluvial system also depends on the period of mining pressure, which is very diverse for different catchments: in some, it is over a hundred years; in others, it is several decades; and in some, it has been completed.

Among the many conditions affecting the nature of matter circulation in mining-industrial and urban areas, the ones of the seemingly utmost importance were characterized.

5.1 Changes of Physicochemical Properties of the Transported Matter

In a substantial part of the Upper Silesian Coal Basin, changes in the physicochemical characteristics of landforms are so insignificant that they did not cause any effect on conditions of matter circulation. In mining areas, where such changes have occurred, the soil is examined within the framework of the planned land reclamation work in order to identify the conditions to introduce vegetation. Changes in the physicochemical properties of such soils may be considered in two aspects. The first concerns the changes in the physical characteristics of ground in the area of mining impact, resulting mainly from changes in moisture. Factors affecting this property include precipitation, the degree of plant cover, lithological features of the substrate, the type of development, etc. and also the desiccation of ground in the areas of depression cones and its hydration in subsidence troughs. In the area of mining subsidence under favourable conditions in the geomorphological and lithological substrate, more or less permanent flooding and ground water inflow take place (Fig. 5.2). The consequence is a reduction of many of the compounds, such as sulphate, phosphate, and nitrate and the destruction of the crumb texture of soil by washing out soil aggregates and closure of larger pores. After drying, the ground is less susceptible to washing out; however, due to its appearance in the generally flat-bottomed subsidence troughs, it does not matter geomorphologically but is important for infiltration, which is difficult without the mechanical loosening of the ground. Soils within the range of depression cones created by the drainage of open-pit and underground mines may undergo long-term overdrying. For example, Banaś et al. (1989) described this type of ground draining in the Ziemowit and Piast mines. In areas of hydrogeological windows in the clay Miocene overburden, mining drainage reaches the ground surface; rainwater and Quaternary-level water are infiltrated intensively to the cracked or porous Triassic and Carboniferous rocks. The draining and over-drying of ground covered almost the entire area of the Lędziny Hills (horst and graben hills with relative heights of about 40 m locally, with clay cover on the slopes) and a part of the Bieruń Hummock. Erosion processes in this area became weakened, as expected. According to Brodowski's research (2009), the drying of the superficial soil horizon significantly reduces the value of the washout because of its crusting; thereby, it reduces its susceptibility to water erosion.

The other aspect of the issue of changes in the physicochemical characteristics of the transported matter is associated with changes in "native" soil characteristics caused by the emergence of new soil, then its mixing with "native" soils (Ostrowski

Fig. 5.2 Water-logged ground in the subsidence trough in the Bobrek valley (Dulias 2003)

2001). The mixing of natural soil with anthropogenic soils may reach a depth of several meters, less frequently over a dozen (Atlas geologiczno-inżynierski 2005; Kupka et al. 2008). In general, changes in the characteristics of landforms related to mining are present in the areas of landfills of the mining industry (nearly 50 km^2), filling subsidence troughs and former excavations with waste material, levelling the surface for urban and industrial development with this material. They are also present along riverbeds in the form of mine tailing embankments and railroad embankments. Waste material, depending on the degree of weathering and vulnerability to leaching and blowing, is transported from its place of deposit by various distances as a result of surface flow, from several to 200–300 m. Due to aeolian transport, the distance might be several kilometres (Szczypek and Wach 1991a, b); it is the longest in the case of river transport at several dozen or more kilometres (Działoszyńska-Wawrzkiewicz 2008) (Fig. 5.3). The subsurface migration of solutions and particulates is complicated: it includes not only midsoil flow, underground runoff, and movement of material through piping, but also flow through excavations, galleries, mining shafts, fissures, and subsidence forms.

The total area of anthropogenic ground is difficult to estimate. Czaja (1992) reported that in 1985 in the Upper Silesian Industrial District, the built-up, industrial, and transportation areas amounted to 310 km^2, which is 270 km^2 more than in 1860 in the pre-mining period. Anthropogenic soils associated with mining are present in the area of about 100–120 km^2.

Fig. 5.3 River transport of coal mining wastes to the Rogoźnik Lake (Dulias 2003)

The area of the USCB is dominated by waste from coal mining, accounting for about 80 % of the total amount of waste (Fig. 5.4). Its mass constitutes mostly tailings (94 %) and only 6 % from extraction. Petrographic and chemical

Fig. 5.4 Vast spoil tip at Skrzyszów in Mszana, the Rybnik Plateau (Dulias 2005)

composition of the latter is varied, with domination of clay rocks (claystones and shales with a high content of kaolinite; Cebulak and Kozłowski 1978). Depending on the mineral composition and the degree of diagenesis, they differ in plasticity, blurring, and swelling (Rosik-Dulewska 2006). Mudstones may constitute up to 40 % of the extractive waste and sandstones; up to a dozen percent consist mainly of quartz sandstone and arkose sandstone with variable mechanical strength and resistance to blurring in water. Qualitative parameters of the tailings are less diverse. They are divided, depending on the fraction, into coarse-grained (20–200 mm), fine (1–20 mm), and flotation (less than 1 mm). In coal mining tailings, the dominating components are SiO_2 and Al_2O_3 (90 %), along with Fe_2O_3 (1–8 %), alkalis (1–4 %), and soluble salts, but mainly chlorides and sulphates (0.2–0.4 %) (Konstantynowicz 1989).

The granulometric composition of newly formed spoil tips includes mostly rocky material; in old, weathered, and burnt heaps, its amount decreases in favour of fine earth (Dwucet et al. 1992). As a result of thermal processes, the pH level of the waste material undergoes a change: at a temperature of 100–300 °C, it decreases to pH 3; at a higher temperature, it increases to neutral or even alkaline (Greszta and Morawski 1972). Burnt spoil tips are also characterized by high porosity. Coal mining waste has a very high content of salts soluble in water. In the case of strongly advancing processes of weathering and erosion, new batches of toxic material are exposed and the leaching processes cause the migration of toxic compounds and their accumulation in the foreland of heaps (Wrona 1975).

Waste stored on tips is prone to natural weathering. Carbon shales disintegrate within a few months and waste-mantle is washed away; claystone shales take longer, and sandstone even longer (Maciak 2003). Mechanical weathering is moreover more intense on unburned spoil tips. The degree of weathering of waste material is determined by the degree of its vulnerability to erosion and denudation processes. Around all observed spoil tips with very poor or partial plant cover, there are proluvial fans at the outlet of erosion furrows cleaving the slopes of these forms (Fig. 5.5).

Zinc and lead ore mining is related to three types of waste: mining, post-flotation, and flotation. The first one is composed of coarse dolomite waste with an admixture of limestone, and, to a lesser extent, clays and sands containing the compounds Zn, Pb, Fe, and S. This type of waste constitutes 25–33 % of the processed ore (Rosik-Dulewska 2006). The material collected on mining heaps is alkaline, characterized by high porosity; the weathering products are easily washed off (Maciak 2003). Post-flotation waste is sludge, containing a mixture of waste from blend and calamine flushers; it is alkaline and contains small quantities of zinc and lead, although this is not the rule: post-flotation waste discharge in the Orzeł Biały Mining and Metallurgical Kombinat in Piekary Śląskie–Dąbrówka Wielka is composed of sludge of granulation up to 10 mm containing 6 % zinc and 1.5 % lead (Janecka et al. 2009). About two-thirds of flotation waste consists of fine material (less than 1 mm) and its composition is dominated by dolomite (70 %) and calcite and partly kaolinite; it is characterized by a high content of heavy metals, which constitutes a threat to the environment in the vicinity of the spoil tips as flotation

Fig. 5.5 Morphological effects of rainwash on the spoil tip in the vicinity of Siersza in the Biskupi Bór Basin (Dulias 2004)

waste dusts after drying. Historical regions of ore mining are characterized by a high degree of heavy metal and sulphate concentration, which is transferred to the natural environment even after 100 years following the ceasing of operations (Cabała and Sutkowska 2006).

In the area of research, there are practically no spoil tips derived from iron ore mining, but there are heaps associated with iron metallurgy and metallurgical energy. They mostly accumulate alkaline steel slag that undergoes hydration in atmospheric conditions. The waste that poses a threat of secondary dusting is silica dust. Other types of waste collected on heaps are forge slag, iron-bearing sludge, and flush water sludge or spent molding sand (Strzelczuk 1977).

Waste from coal-powered plants consists of slag and furnace dust with different chemical composition and physical properties. The silica content in flue dust is about 47 %; in slag, it is up to 65 %. On hot days, ashes on the heaps heat up to 45 °C, whereas at night they cool down quickly to 10 °C (Dwucet et al. 1992). Due to significant participation of dust fraction, the material is susceptible to blowing out.

Some of the older and smaller-volume spoil tips in the Upper Silesian Coal Basin have already been largely washed by precipitation (Wilk 2003). The leaching process of toxic substances presently takes place on a smaller scale than in the case of new spoil tips (Szczepańska 1987). Infiltration leaching of chloride from heaps of a dozen meters in height may last several years; for sulphate, it lasts at least several decades. Ground with anthropogenically altered physicochemical characteristics is present above all areas of the most intensive mining and in urban areas on the Bytom, Katowice, and Rybnik Plateaus.

5.2 Sediment Production Zone

Slopes supply rivers with matter. The defining characteristics of the productive potential in this area primarily include the catchment area, the inclination of slopes, relative heights, density of valleys, and also the type of sediment and land use. Changes in the above-mentioned characteristics, may, due to mining, significantly disrupt the functioning of not only the slope but also the channel subsystem, which is sensitive to events on slopes (Mazurek and Zwoliński 2012).

Changes in watersheds. One of the effects of mining activities in the USCB are changes in the course of watersheds and resulting changes in the area of catchments. The problem has been signalled from the mid-twentieth century. Czaja (1988), based on analysis of cartographic materials from the nineteenth and twentieth centuries, presented changes in the Vistula-Oder watershed within the Upper Silesian Industrial Region; the author showed that in a 100 years, its position has changed in 75 % of its length due to surface subsidence of land and urban sewer systems installed in the watershed area. Other causes of changes in watersheds include the creation of spoil tips and the construction of high rail and road embankments. Changes in watersheds occur in catchments of various orders. Figure 5.6 shows changes in a watershed of the first order (the Vistula, the Oder within the Bytom Plateau) and changes in a watershed of a small basin of the sixth order (the Podkościele watercourse, the Nacyna catchment on the Rybnik Plateau), both in the period of 1883–1993.

An area of 122.3 km^2 was excluded from the existing fluvial system; therefore, in many catchments, the area drained by surface flow significantly decreased. An extreme example is the Upper Rawa catchment located in the Carboniferous zone, which was excluded from surface flow by almost 97 % (Table 5.2). In some catchments located in the Siemianowice Upland in the Triassic zone, not less than 70 % of their area is drained and a specific case is observed in the Michałkowice Ditch catchment, where landlocked basins occupy an area of over 20 km^2. In the Miocene zone, the biggest changes in watersheds took place in the Dębinka catchment, where almost half of its surface is currently excluded from fluvial drainage. Changes in the areas of some catchments are so large that they are most likely to be reflected in the course of erosion and denudation.

Changes in slope inclination. In the studied catchments, changes in slope inclination are similar in nature to those presented in Sect. 4.3 in relation to geomorphological units. Their most significant feature is the reduction of the participation of flat areas observed in as many as 90 % of the studied catchments. The catchment of the Kłodnica (within the study area) experienced a loss of 33 km^2; the catchment of the Przemsza experienced a loss of 46 km^2; and Bierawka experienced

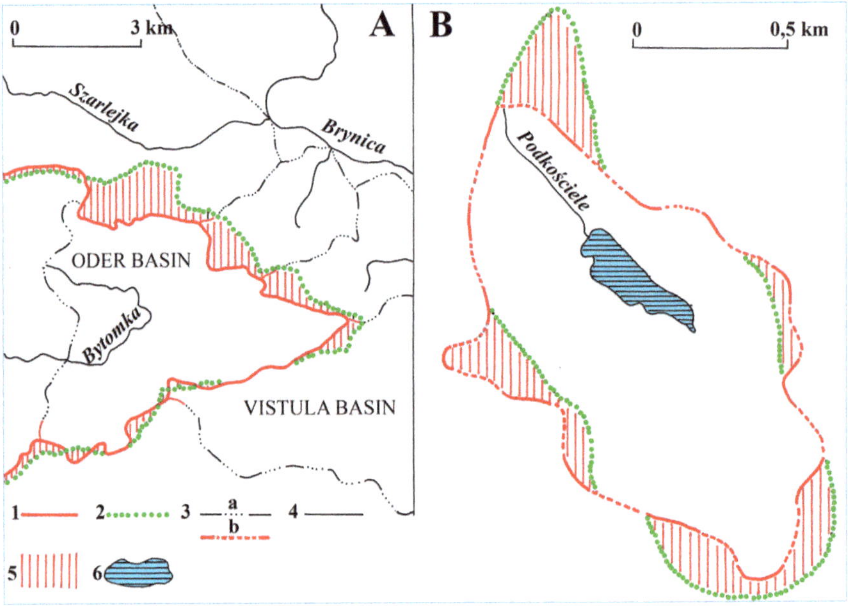

Fig. 5.6 Changes of watersheds in the years 1883–1993 for (**A**) the first order Vistula-Odra on the Bytom Plateau and (**B**) the sixth order Podkościele on the Rybnik Plateau. (**1**) watershed of the first order in 1993, (**2**) watersheds in 1883, (**3**) watersheds of the lower orders in 1993: fourth (**a**) and sixth (**b**) order, (**4**) rivers and watercourses, (**5**) zone of changes in the catchment area, (**6**) water reservoir

a loss of 11 km^2. The biggest loss of plains in favour of slopes with an inclination of 1°–3° (and rarely 3°–5°) was observed for smaller catchments located in the Miocene and Triassic zones. Only in 6 catchments did the flat areas increase, and all of those are located in Triassic and Carboniferous zones. Areas with the largest increase of plains included the catchments of the Miechowice Stream (+11 %), the Michałkowice Ditch (+11 %) and the South Bolina 1 (+8.3 %) (Table 5.3).

The participation of highly inclined (9°–20°) and steep (above 20°) slopes increased slightly (in less than 1 % of the slopes). The main cause of this was the creation of large, artificially formed spoil tips with steep slopes and in catchments with naturally large slope inclinations—the "transition" of moderately inclined slopes to the category of highly inclined slopes as a result of surface subsidence. Catchments that observed the largest increase of these slopes included the Radlin Stream (+4.3 %) and the Orzeł Biały Ditch (+3.2 %).

In the period of 1883–1993, a total of 94 % of catchments reported an increase in an average slope inclination by 0.4°, up to as much as 1.8° in the case of the Orzeł Biały Ditch catchment (the Siemianowice Upland), by 1.3° in the Jordanek catchment (the Podstokowa Zone and the Mikołów Hummock) and 0.9° in the catchment of the Murcki Ditch (the Murcki Plateau) and the Radlin Stream

Table 5.2 Landlocked areas in selected catchments in the Upper Silesian Coal Basin according to geological zones (from Dulias 2013)

Basin	Catchment	Catchment area (km^2)	Landlocked areas	
			Area (km^2)	Percentage of the catchment area
Carboniferous zone				
Rawa	Upper Rawa	5.36	5.19	96.8
	Open Rawa Channel	46.42	7.00	15.1
Triassic zone				
Szarlejka	Segiet	19.79	5.54	28.0
	Michałkowice Ditch	23.84	20.54	86.2
Brynica	Dąbrówka Wielka Ditch	9.23	4.43	48.0
	Orzeł Biały Ditch	6.37	3.67	57.6
	Upper Bytomka	17.99	7.83	43.5
Bytomka	Miechowice Stream	2.56	0.86	33.6
	Drzymała Ditch	5.26	1.9	36.1
Miocene zone				
Pszczynka	Upper Pszczynka	8.72	1.87	21.4
	Dębinka	8.19	4.03	49.2
Ornontowice Stream	Gierałtowice Stream	4.42	0.68	15.4
	Beksza Stream	14.47	2.09	14.4
Bierawka	Knurówka	18.14	2.29	12.6
Nacyna	Tributary from Michałkowice	21.52	1.77	8.2
Szotkówka	Stream from Gogołowa	3.11	0.45	14.5

(the Rybnik Plateau). The above data indicates that in almost all studied catchments, the erosion potential of the slope subsystem clearly increased.

Changes in the relief energy. To determine the changes of relief "energy" in the studied catchments, at least in general terms, the relief indicator L was calculated for them, which is the ratio of the catchment's denivelation to its length (the length of the main valley from the mouth to the watershed). At the end of the nineteenth century, before the intensive development of the mining industry, the relief indicator L spanned a very wide range—from 3.4 to 27.8. The highest values were associated with smaller catchments in highly elevated areas, such as the Murcki Plateau (the Tributary from Wesoła: 27.8; the South Bolina: 2–17.6) and the Rybnik Plateau (the Stream from Gogołowa: 18.8; the Pludry: 16.5), as well as the Miechowice Stream catchment on the southern slope of the Miechowice Upland (22.4). The lowest relief values were obtained for the catchments of major rivers in the Racibórz-Oświęcim Basin—the Kłodnica (3.4), the Nacyna (4.5), the Mleczna (5.3), the Bierawka (6.0)—as well as for smaller catchments, such as the Paniówki Stream (4.7) or the Gierałtowice Stream (6.5).

Table 5.3 Changes in the share plains in selected catchments between 1883 and 1993 against a background of geological zones (from Dulias 2013)

Geomorphological unit	Catchment	Percentage share of the plains in the catchment		Changes in the share plains in the years 1883–1993	
		1883	1993	(%)	(km²)
Carboniferous zone					
Murcki Plateau	Bolina	27.7	30.8	+3.1	+0.87
	South Bolina 1	23.9	32.2	+8.3	+0.60
Podstokowa Zone, Ruda Hills, Kochłowice Hills	Bielszowice Stream	36.7	27.9	−8.8	−2.81
Triassic zone					
Siemianowice Upland	Orzeł Biały Ditch	49.8	23.7	−26.1	−1.66
	Dąbrówka Wielka Ditch	53.5	35.5	−18.0	−1.66
	Michałkowice Ditch	42.5	53.5	+11.0	+2.62
Miechowice Upland	Miechowice Stream	27.9	38.9	+11.0	+0.28
	Tributary in Mikulczyce	32.7	36.3	+3.6	+0.10
Tarnowice Plateau, Miechowice and Siemianowice Uplands	Szarlejka	42.2	31.0	−11.2	−4.68
Miocene zone					
Podstokowa Zone	Stream from Paniówki	86.6	58.7	−27.9	−0.73
Podstokowa Zone, Mikołów Hummock	Jordanek	46.4	33.0	−13.4	−1.38
Rachowice High Plain	Cienka	67.2	47.0	−20.2	−2.32
	Gierałtowice Stream	62.6	49.8	−12.8	−0.57
	Knurówka	78.0	69.6	−8.4	−1.52
Rybnik Plateau	Dębinka	60.8	46.7	−14.1	−1.15
	Tributary from Kąty	33.1	19.8	−13.3	−0.30
	Upper Pszczynka	50.2	37.5	−12.7	−1.11
	Marklówka	35.2	23.1	−12.1	−1.09
	Stream from Gogołowa	29.1	19.4	−9.7	−0.30
	Tributary from Popielów	34.0	24.3	−9.7	−0.91

Changes in relative heights and the length of catchments (e.g. as a result of the creation of landlocked basins in the area of a watershed) that occurred in the twentieth century (until 1993) resulted in an increase of the relief indicator for 80 % of studied catchments by an average of 1.6. The largest (even reaching a doubled *L* ratio) refers to the catchments with vast spoil tips of several dozen metres in relative height, such as the Upper Pszczynka, Szotkówka, and Nacyna (Table 5.4). Catchments with extreme values of the *L* indicator in the pre-mining period (the highest or the lowest), with a few exceptions, increased the energy of the relief to a relatively small extent.

Changes in the drainage density. The catchments of the study area are generally not diverse in terms of the drainage, mostly being 1.4–1.7 km/km^2. Such values were obtained in the catchments of the Bytomka, the Bielszowice Stream, the Czarniawka, the Knurówka, the Jordanek, the Cienka Stream, the Murcki Ditch, the Paniówki Stream, and the Dąbrówka Wielka Ditch. Therefore, catchments are of both large and small areas and are situated in different geological zones and geomorphological units. Against this background, the Rybnik Plateau stands out with its dense network of valleys developed on loess cover. For example, in the catchment of the Pludry stream, the drainage density is 3.6 km/km^2; in the catchment of the Dębinka, it is 2.9 km/km^2; and in the catchment of the Jastrzębianka, it is as high as 5.2 km/km^2. On this basis, it can be inferred that the Rybnik Plateau has extremely favourable conditions for the discharge of matter from the slope subsystem. Changes in the drainage density are relatively small. They mainly result from the backfilling of some small valleys with mining waste or their transformation into landlocked basins.

Changes in the drainage density have not been the subject of detailed analysis. According to Czaja (1997), the drainage density in the catchment of the Bytomka in the years 1860–1994 reduced from 1.3 to 0.53 km/km^2; in the catchment of the

Table 5.4 Catchments with the highest changes in relief energy ratio *L* in the years 1883–1993 (from Dulias 2013)

Catchment	Relief energy ratio *L*		Changes in relief energy ratio *L* in the years 1883–1993
	1883	1993	
Upper Pszczynka	7.7	15.5	+7.8
Szotkówka (to the Jastrzębianka mouth)	7.9	14.9	+7.0
Nacyna	4.5	10.6	+6.1
Jordanek	11.3	16.0	+4.7
Tributary from Przyszowice	9.1	12.6	+3.5
Stream from Gogołowa	18.8	21.7	+2.9
Cienka	7.6	10.1	+2.5
South Bolina 1	14.0	16.2	+2.2
Orzeł Biały Ditch	11.4	13.6	+2.2
Miechowice Stream	22.4	15.2	−7.2

Rawa in the years 1801–1994, it reduced from 0.93 to 0.38 km/km^2. Changes in the length of the river (water) network in a catchment may run in a different direction than the changes in the length of the main river; for example, the main river may be shortened, and the length of the river (water) network in the watershed may also increase due to the emergence of drainage ditches, canals, etc. Czaja observed that the water network was shortened in built-up areas but it was extended in forests and agricultural areas due to the network of drainage ditches in areas subject to mining subsidence.

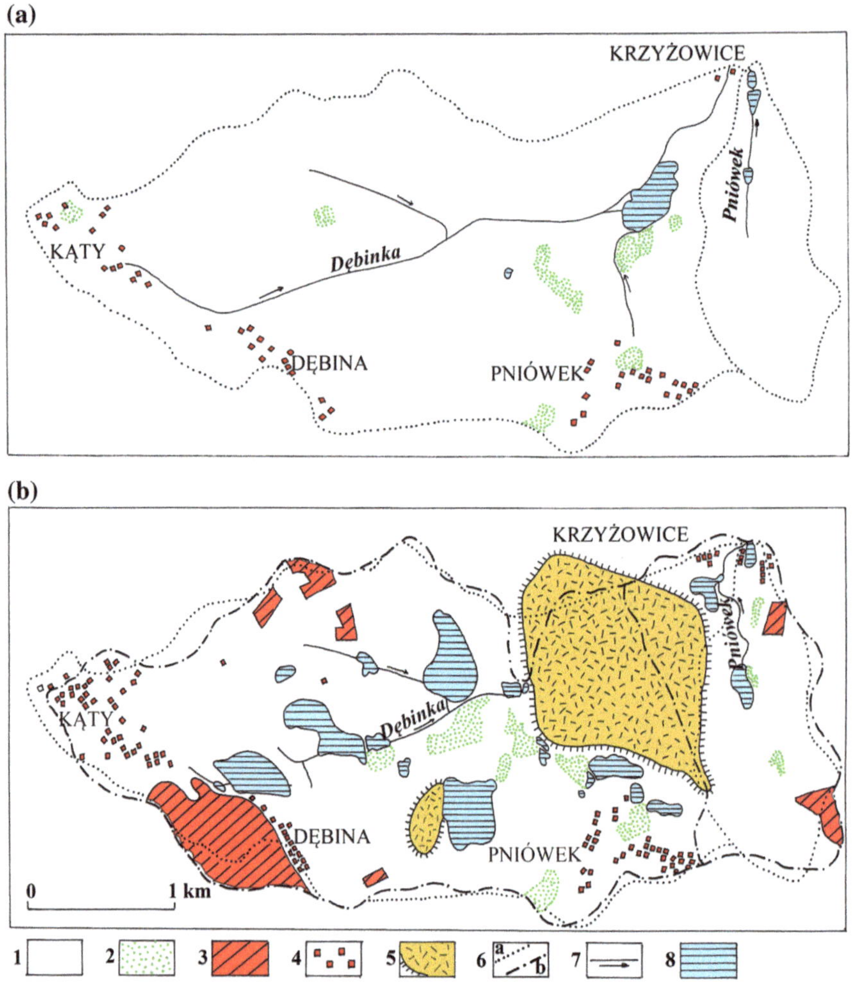

Fig. 5.7 Land use in the Dębinka and Pniówek catchments on the Rybnik Plateau in the years 1960 (**A**) and 2004 (**B**) (based on Dulias 2008b): (**1**) agricultural land, (**2**) forests, (**3**) compact housing, (**4**) scattered housing, (**5**) spoil tips, landfills, (**6**) watersheds in the years 1960 (**a**) and 2004 (**b**), (**7**) watercourses, (**8**) water reservoirs

Changes in land cover and land use. Land cover and its use constitute important features of the area from the point of view of erosion and denudation processes. In the last 100–200 years, the USCB has undergone significant changes in this respect. In the mid-nineteenth century, compact urban and industrial development occupied only 2.2 % of the Upper Silesian Industrial Region; in 1985, the number went up to 22.7 % to 220 km^2 (Czaja 1992). Such a direction of changes in land use caused a reduction in permeable areas in favour of impermeable areas. In the 1970s, the annual loss of rainwater infiltration in the built-up area of the USIR was estimated at about 90 million m^3 (Jankowski 1992).

Changes in land use in catchments took different courses, but they mostly consisted of the conversion of agricultural or forestry land into built-up areas. In some catchments, they represent over 60 % of the area, such as in the Sośnica Stream catchment (the Podstokowa Zone), the Ruptawka (the Rybnik Plateau), and the Nowobytomka (the Ruda-Chorzów Hills). Some catchments have maintained their agricultural character, such as those located on the Rachowice High Plain—the Tributary from Przyszowice and the Cienka Stream, where agricultural land still occupies more than 80 % of the area.

In some previously agricultural catchments, water reservoirs and spoil tips were formed in subsidence troughs. For example, in the Dębinka and Pniówek catchments on the Rybnik Plateau, until the 1960s more than 80 % of the land was arable; at present, it is less than 50 %, with over 15 % taken up by spoil tips and almost 8 % by water reservoirs in subsidence troughs (Dulias 2008b) (Fig. 5.7). Typical forest catchments (over 80 % of the area) are represented by the Murcki Ditch and South Bolina 1, located on the Murcki Plateau.

5.3 Transfer Zone

Changes within riverbeds in the studied mining area may be divided into two groups. The first involves physical changes in the characteristics of the bed, such as the length, gradient, or course. The other consists of changes in the amount of water and matter transported in the bed. The main reasons for these changes include mining activity and the intensive urbanization of the USCB and related hydro-technical work forced by their effects or needs.

Changes in river courses and hydro-technical development. The most noticeable changes in river beds are those associated with their geometry—the straightening of their course, repaving of bed segments, deepening, the construction of drops, the strengthening of the bottom and edges, the river banks, etc. (Fig. 5.8). At the beginning of the twentieth century, more than 90 % of watercourses in the Upper Silesian conurbation flowed in natural channels; in 1994, this number was only 34 %. Hydrotechnical works were commenced in 1928 with the regulation of the Rawa and its tributaries, and they continued after World War II, starting with the Brynica. In addition to the strengthening of their banks with cobbles or fascine, some river beds were put in stone and concrete troughs (mostly in their lower or

Fig. 5.8 Nowy Dwór watercourse, with paved banks and deepening of the channel to dewater flooded plains in the middle section of the valley (Dulias 2006)

middle segments), such as the Rawa, the Biała Przemsza, the Przemsza, the Kłodnica, the Goławiec Stream, the Gostynia, the Tychy Stream, and the Stawowy. The Brynica riverbed was completely sealed (within the study area), which aimed to reduce water escaping into the karstic Triassic bedrock. In many rivers and streams, concrete or stone drops were built to reduce the longitudinal decline (e.g. the Bierawka, Kłodnica, Bytomka, Szotkówka). The rivers that have clearly changed their morphological character due to the straightening of their curving or meandering segments include the Brynica, Szarlejka, Rawa, and Kłodnica.

In the study area (2,838 km^2), the length of embankments along riverbeds, according to the measurements of topographic maps of 1: 10,000 (1993), is 300 km. The length and height of the embankments is constantly changing. For example, the bed of the Szotkówka in the section in Połomia, in the early 1990s, was not yet embanked (Mapa topograficzna 1993). In 2005, it was lined with mine tailing embankments several meters in height, which are now within the water reservoir in the subsidence trough and are barely visible from the water surface.

In areas with nondiversified relief, even a relatively small subsidence of the surface forces the construction of embankments along riverbeds. An example is the lower sections of the Pszczynka and Korzenica within the activities of the Czeczott mine. Here, the subsidence of the surface by about 2.5 m resulted in the flooding of the area and the need to displace more than half-kilometre long sections of both rivers and their embankments at a length of about 2 km in the early 1990s. The costs of damage removal in river valleys and areas of overflow represent a significant proportion of the costs that mines have to bear for the liquidation of mining damages. In the period of 1996–2000, it was 16 %, which is about PLN 0.5/tonne of extracted coal mined (Kaszowska 2005).

For many years, one of the major hydro-technical problems in the USCB has been the regulation of the water regime in the valleys of the Kłodnica and its tributaries in the district of Zabrze–Makoszowy (Fig. 5.9). The area is located within intensive mining subsidence; due to small denivelations and a shallow water table, it is particularly susceptible to water inflows and a frequent occurrence is the reverse of riverbed gradient and backwater. Mine tailing embankments here require frequent raising for flood protection (Wach and Szczypek 1996). A high risk of flooding is also present in the lower part of the Gostynia. Most of the embankments occur along the Vistula catchment (219 km), the longest accompany riverbeds of the Brynica (41 km), the Mleczna (32 km), the Czarna Przemsza (30 km), and the Gostynia (28 km).

Changes in river lengths. The results of a detailed comparative analysis of contemporary maps with maps of the late nineteenth century indicate that out of the 48 studied rivers, 3 did not change their lengths, 23 rivers were shortened, and 22 were extended. The shortening mostly affected large rivers (Table 5.5), mainly due to the straightening of their middle and lower sections and the decline of their upper sections (e.g., the Bielszowice Stream and the Szotkówka). Several rivers shortened, although their upper sections were significantly extended—in the case of the

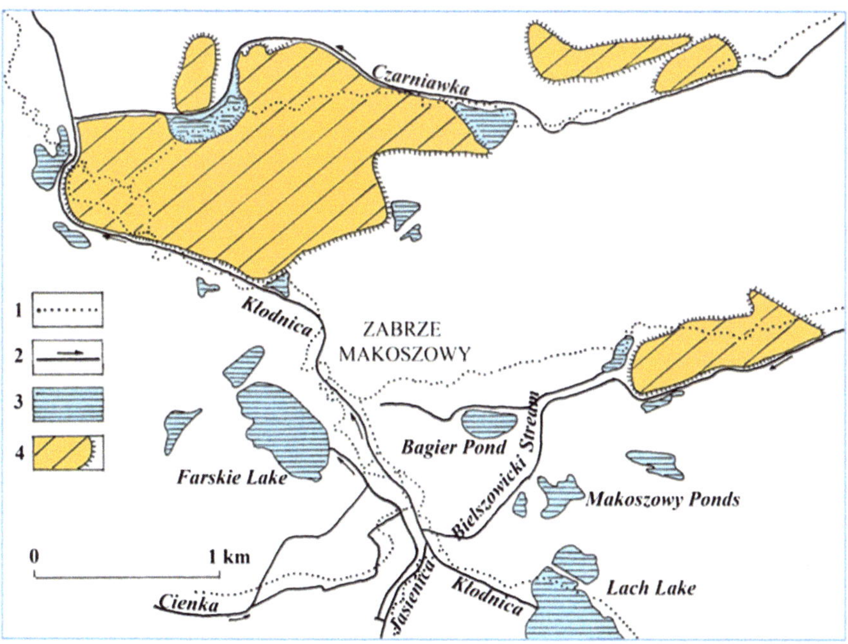

Fig. 5.9 Changes of the Kłodnica river course and its tributaries in the years 1883–1993: (**1**) river network in 1883, (**2**) river network in 1993, (**3**) main water reservoirs in 1993, (**4**) spoil tips

Table 5.5 Changes of lengths and gradients of large rivers in the area of the Upper Silesian Coal Basin in the years 1883–1993 (from Dulias 2013)

River	Length (km)		Average gradient (‰)	
	1883	1993	1883	1993
Oder Basin				
Kłodnica (from Kościuszki Street in Katowice to the sluice in Gliwice)	37.9	35.7	1.8	1.8
Bierawka (to the Knurówka mouth)	21.2	20.4	3.5	4.0
Bytomka	22.5	22.0	2.8	2.9
Bielszowice Stream	14.8	14.3	3.6	4.3
Nacyna	19.1	17.9	3.2	4.3
Vistula Basin				
Przemsza (from the Biała Przemsza mouth to the Imielinka mouth)	18.6	18.6	1.0	1.0
Brynica (from the Szarlejka mouth to the confluence with the Czarna Przemsza)	37.3	22.6	0.6	1.1
Rawa (from the beginning of the Open Rawa Channel to the confluence with the Brynica)	16.6	15.9	1.6	1.8
Mleczna	23.8	22.3	2.0	2.1

Nacyna, it was even up to 1.7 km, but it generally shortened by 1.2 km. Medium and small watercourses underwent lengthening. The main reason for this, with the exception of 3 cases, was an increase in the length of the upper sections, such as for the Jordanek, the Szczygłowice Stream, the Knurówka, the Marklówka and the Tributary in Mikulczyce; these are mostly watercourses located in the Miocene zone in the Oder basin (73 %). Examples located in the Carboniferous zone include the Czarniawka, the Bolina, and the Leśny Stream. The Vistula basin decreased by almost 20 km of rivers (11 % of their length in 1883), whereas the Oder river basin increased by 7 km (3 % of their length in 1883). These results relate to 430 km of rivers covered by the analysis.

Changes to the base level of erosion. In areas shaped by flowing water, it is not possible to lower the bottom of the valley below the base level of erosion, with the exception of areas influenced by glacial and aeolian processes (Migoń 2006). The Upper Silesian Coal Basin example shows that these exceptions also include mining areas located within the range of intensive mining subsidence. In many places, land adjacent to the riverbed is located a few meters below its bottom. Underground mining activity is, in fact, carried out irrespective of the hypsometric relationships on the surface. The resulting surface deformations include various part of catchments: watershed areas, slopes, and valley bottoms. If mining subsidence occurs at the river mouth, then the base level of erosion is lowered. If they include watershed areas, their parts may be excluded from the catchment by transforming into landlocked basins.

In 1993, only two of the studied rivers had an increased base level of erosion, compared to the end of the nineteenth century: the Nacyna (4 m), due to the establishment of the Rybnik Reservoir, and to a small extent, the Murcki Ditch (0.5 m). In the case of 6 rivers, their base level of erosion did not change. Except for the Przemsza and the Ślepiotka, these are small watercourses (3–5 km long). For most rivers (82 %), their base level of erosion was lowered as a result of mining subsidence: in the Vistula basin, it was lowered by an average of 3.4 m; in the Oder basin, it was lowered by 4.8 m. The lowering of the base level of erosion mostly affected rivers of the Miocene zone, in the Rachowice High Plain, in the Podstokowa Zone, and the Rybnik Plateau (Table 5.6).

Table 5.6 Rivers with the greatest lowering of the base level of erosion in the years 1883–1993 (from Dulias 2013)

River/stream	Geomorphological unit	Changes in base-level of erosion (m)	Changes in river gradient (‰)	Changes in river length (km)
Oder Basin				
Cienka	Rachowice High Plain	−13.2	+3.0	+1.1
Tributary from Przyszowice	Rachowice High Plain	−10.5	+2.6	+0.5
Czarniawka	Podstokowa Zone, Ruda Hills	−8.0	+0.6	+1.5
Bielszowice Stream	Podstokowa Zone, Ruda Hills, Kochłowice Hills	−7.0	+1.7	−0.5
Jordanek	Podstokowa Zone, Mikołów Hummock	−7.0	+2.5	+0.1
Szczygłowice Stream	Rachowice High Plain, Podstokowa Zone	−6.0	+2.8	+0.4
Knurów Stream	Rachowice High Plain	−6.0	+2.4	+1.6
Tributary from Popielów	Rybnik Plateau	−6.5	+2.3	+1.0
Szotkówka (to the Jastrzębianka mouth)	Rybnik Plateau	−5.6	+1.8	−0.4
Jastrzębianka	Rybnik Plateau	−5.6	+1.8	−0.3
Radlin Stream	Rybnik Plateau	−5.2	+4.4	+2.3
Vistula Basin				
Dąbrówka Wielka Ditch	Siemianowice Upland	−6.0	−0.2	+0.6
Bolina	Murcki Plateau	−5.5	+0.1	+0.3
South Bolina 1	Murcki Plateau	−7.0	+1.8	+0.3
South Bolina 2	Murcki Plateau	−5.9	+0.9	–
Goławiec Stream	Chrzanów Graben, Lędziny Hills	−4.5	+0.4	−0.9
Radzionków Dich	Tarnowice Plateau	−4.4	+2.0	−1.7

Fig. 5.10 Example of increasing river erosion as a result of surface subsidence in a tributary of the Kłodnica River near Paniówki, the Podstokowa Zone (Dulias 2012)

The consequence of the lowering in base level of erosion of rivers is an increase in the rate of their erosion, which was observed in the cases of the Kłodnica (Wach 1987b), the Bolina (Molenda 1999), or the Stream from Gogołowa (Mackiewicz et al. 1979; Jankowski 1986) (Fig. 5.10). In the years 2006–2012, observations of the lower section of the Kolejówka on the Rybnik Plateau were carried out. In 2006, the riverbed floor and edges were lined with concrete bricks. On the threshold with the height of 0.7 m, occurring within several meters from the mouth of the Kolejówka into the Szotkówka, the bricks were heavily destroyed, while the mouth was hanged over the Szotkówka's riverbed by about 1.5 m and limited with very steep, 3-m-high edges. In 2012, in the same segment of the riverbed, there were no thresholds, the concrete strengthening was destroyed in 90 %, the bed was clearly widened, and the slopes were gentler (Dulias 2013). The effects of the particularly high river erosion caused by the lowering of the base level erosion was observed in the valleys of the Ornontowice Stream, the Czarniawka, and the Jasienica.

Changes in river gradients. Mining subsidence causes deformation of river valleys, leading to changes in river gradients: their increase, decrease, or reversal. An increased gradient leads to the acceleration of erosion processes, and as a consequence, the reduction of water speed and increased accumulation of the carried material. The riverbed becomes shallower; it can also lead to the accumulation of water and flooding of areas outside the riverbed (Jankowski 1986; Czaja 1999). The flattening of the longitudinal profile of rivers, combined with an increase in the amount of material carried by rivers (particularly the suspension) speeds up the build-up of floodplains (Wach 1987a, b). The passage of the exploitation front under the bed of the watercourse, especially in its middle section, may also lead to a reversal of the gradient and the emergence of a floodplain, with the water table reaching the ordinate of this reversal (Eckes and Żuławski 1988; Mackiewicz et al. 1979; Pozzi et al. 2008). These are sedimentary basins for sediments carried by rivers. Rivers in the areas of mining subsidence are often characterized by the presence of long sections of insignificant river gradients alternating with short sections with high gradients (e.g. Jankowski 1986; Wach 1987a, 1991; Eckes and Żuławski 1988; Czaja 1999).

Research results obtained for 48 rivers and watercourses in the area of the USCB show that the increase in river gradient is common: it was observed for 76 % of rivers, including almost all major rivers, except the Przemsza and the Kłodnica. In the case of the Kłodnica, despite the fact that an average gradient in 1993 was identical to the decline in 1883, there are significant differences in gradients in particular parts of the river. In the segment between the mouth of the Jasienica and the mouth of the Ślepiotka, the river gradient increased by 0.6 %. Between the mouth of the Jasienica and the mouth of the Bytomka, the gradient was only 0.3 %, which is more than 3 times lower than in the late nineteenth century (1.1 %). In the upper course, up to the mouth of the Ślepiotka, the gradient was 6.4–0.9 % lower than in 1883.

For the Brynica, which originates outside the USCB, its segment from the Szarlejka mouth to the confluence of the Brynica with the Czarna Przemsza was examined. In the period of 1883–1993, the river fundamentally changed its morphological character. Instead of the natural meandering riverbed, an artificial, straight, paved, and embanked channel was created. The length of the river in the analysed segment decreased by 14.7 km, which is almost 40 %. No other river in the study area has been transformed to such an extent. The average river gradient of the Brynica almost doubled—from 0.6 to 1.1 %. In the course of the river, there are segments that distinctly differ in gradient (twofold, threefold), both between one another and also in relation to the drop of 1883.

A significant increase in the average river gradient was recorded in the case of large rivers, such as the Bielszowice Stream (+0.7 %), the Nacyna (+1.1 %), and the Bierawka (+0.5 %). However, it mostly increased in the case of small and medium rivers. The data listed in Table 5.6 show that an increase in the river gradient is associated with a lowering of the base level of erosion; simultaneously, it is not directly dependent on the shortening of rivers because most of them extended. All watercourses with average river gradients increased by more than 2 % and are located in the Oder basin, almost entirely in the Miocene zone; already in the pre-mining period, they were characterized by significant gradients. For example, an average river gradient of the tributary from Kąty was 8.8 % in 1883 and 11.4 % in 1993 (+2.6 %), the Radlin Stream was 6.4 % in 1883 and 10.8 % in 1993 (+4.4 %), and the Jordanek was 7.9 % in 1883 and 10.4 % in 1993 (+2 5 %).

A decrease in an average river gradient was observed for 10 out of the 48 studied watercourses. They mostly belong to the Vistula basin (7) and constitute short watercourses with an average length of 3.8 km (1993). During the pre-mining period, they were characterized by large and very large gradients amounting to an average rate of 9.2 %, but the values in their broader segment ranged from 4.1 % to even 20.4 % for the Miechowice Stream, which is the maximum value recorded for the studied watercourses. The reduction of the gradient is significant, with an average of 1.8 % and a maximum of 3.9 % for the Orzeł Biały Ditch. This watercourse stands out from the others: in the pre-mining period, it had a gradient of 4.1 %; however, due to very large surface subsidence in the heads of a river area (11 m), the gradient decreased to only 0.2 %. Most of these watercourses (7) have slightly reduced lengths, which once again indicates that in the area of mining subsidence there is not straight dependency between the change in the length of a watercourse and the change of its gradient. However, the lowering of the absolute altitude of the heads of the studied rivers—on average by 8.1 m and by a maximum of 18 m—had an effect on reducing the gradient.

Overall, in the period of 1883–1993, an average drop of rivers draining the research area into the Vistula decreased from 5.9 to 5.6 %. For the rivers belonging in the Oder basin, it increased from 6.8 to 7.6 %. This means that natural differences in the erosion potential of the two drainage basins significantly increased as a result of anthropological pressure, and especially mining.

In the areas of mining subsidence, noticeable changes affect the shapes of longitudinal profiles of rivers (Posyłek and Rogoż 1982; Jankowski 1986; Wach 1987a, 1991; Dulias 2006, 2008a, 2011). They have different characteristics, as deformities may include the entire valley or its individual parts, also with diverse or the same intensity (Fig. 5.11). A significant lowering in the upper part of the valley

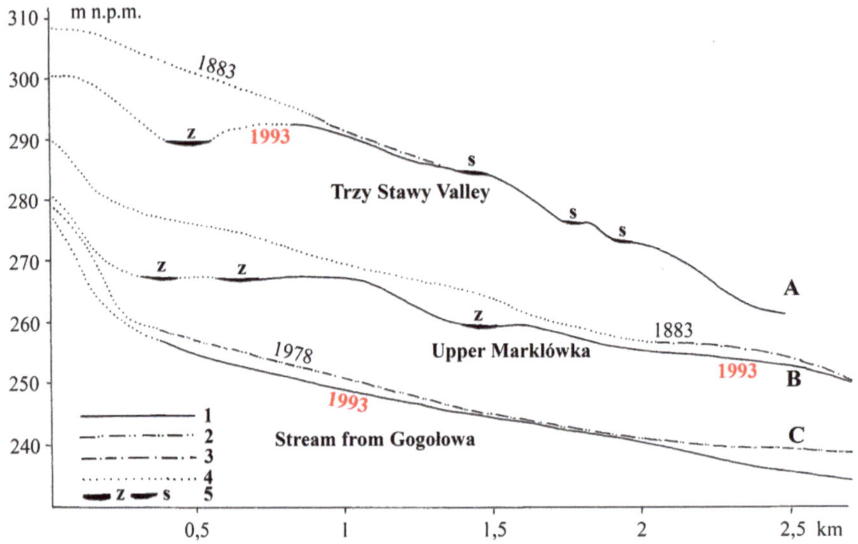

Fig. 5.11 Changes in the longitudinal profiles of selected watercourses resulting from mining subsidence (from Dulias 2013): (**A**) the Trzy Stawy Valley, a tributary of the Rokitnicki Stream, the Miechowice Upland; (**B**) the upper section of the Marklówka, a tributary of the Lesznica River, the Rybnik Plateau; (**C**) the Gogołowa Stream, a tributary of the Szotkówka River, the Rybnik Plateau; Longitudinal profiles in the years: (**1**) 1993, (**2**) 1978, (**3**) 1883; (**4**) longitudinal profiles of dry sections of valleys; (**5**) water reservoirs: *z* in subsidence basins, *s* ponds

may cause its separation as a land-locked basin, thereby leading to the shortening of the valley; cutting the level of groundwater by the subsiding surface may affect the transformation of the previously dry valleys into valleys with fixed watercourses. Examples from the area of the Rybnik Plateau are the upper Marklówka and the valley of the Rydułtowy Stream in which, only since the mid-twentieth century as a result of mining subsidence by about 5–8 m, a watercourse of over 1.4 km in length emerged (Dulias 2008a, 2011).

Changes in river flows. The location of research in the watershed area of the Vistula-Oder makes it poor in water resources. The vast demand for water for industrial and municipal purposes imposes the necessity of its transfer from catchments located outside the research area. The local water cycle also includes large amounts of water from mine drainage: in 1850, 3–4 m^3/s; in 1900, 6–6.5 m^3/s; in 1955, 10–12 m^3/s; in the 1970s and 1980s, about 20 m^3/s; and in 1994, 16 m^3/s (Czaja 1999). The large participation of "alien" water in the water cycle is a characteristic feature of the hydrological regime of rivers in the study area. Consequently, minimum flows are clearly increased, runoff within the year is more uniform, and an average runoff moduli reaches high values (Jankowski 1986, 1987; Czaja 1999).

In the 1980s, the size of water transfer into the province of Katowice matched the flow of 10–11 m^3/s. The supply of water discharged from coal mines, zinc and lead ores, and stowing sands (in the decade from 1970 to 1979) was 1.8 million m^3/day, corresponding to a flow of 20.8 m^3/s; from this, 10.7 m^3/s flowed directly into rivers. The river outflow therefore had almost 22 m^3/s of "alien" water —that is, more than 30 % of the average outflow, estimated at 63–65 m^3/s (Jankowski 1987, 1992; Konstantynowicz 1989). The anthropogenic outflow is particularly large during average low flows (Table 5.7). According to Czaja (1999), in the Brynica in the periods of medium and low flows, natural water accounted for only 2–3 % of the river flow. An example of a river with a very anthropogenic nature is the Rawa, which carried 59 % "alien" water in the early twentieth century (1912) and 92 % in 1980 (Jankowski 1988). Rivers with large, but a relatively smaller, participation of "alien" water in the river outflow include the Bierawka, the Nacyna, and the Szotkówka, with the average annual outflow amounts to 16–24 % (Jankowski 1986). The Mleczna carried from 31 to 62 % "alien" water in the period

Table 5.7 The share of "alien" waters in the medium-low runoff of selected rivers in the area of Upper Silesian Coal Basin (after Sołtysik 1980; Jankowski 1987)

Oder Basin		Vistula Basin	
River	Percentage share of alien waters	River	Percentage share of alien waters
Kłodnica—above the mouth of the Bytomka	57.9	Biała Przemsza—Maczki	53.0
Bielszowice Stream—mouth to the Kłodnica	81.5	Brynica—Sosnowiec	68.5
Żernica Stream—mouth to the Bytomka	70.4	Bolina—mouth to the Czarna Przemsza	83.1

Table 5.8 Average and maximum water levels and flows in large rivers in the Upper Silesian Coal Basin with the corresponding runoff moduli (after Mapa Hydrograficzna 2001–2003)

River	Profile	Water level (m)		Flow (m^3/s)		Runoff moduli ($dm^3/s/km^2$)	
		Average	Maximum	Average	Maximum	Average	Maximum
Vistula Basin							
Brynica	Szabelnia	0.46	1.50	5.70	41.9	11.80	86.8
Czarna Przemsza	Radocha	0.57	2.12	4.56	64.0	8.76	123.0
Biała Przemsza	Niwka	2.27	3.76	7.50	61.2	8.56	69.9
Przemsza	Jeleń	2.20	5.98	19.70	105.0	9.87	52.6
Bobrek	Niwka	2.56	3.60	1.25	27.6	10.50	232.0
Gostynia	Bojszowy	0.79	3.96	3.49	62.6	10.50	189.0
Mleczna	Bieruń Stary	0.45	2.22	1.38	43.6	11.30	358.0
Oder Basin							
Kłodnica	Gliwice	1.02	4.05	6.41	88.1	14.40	198.0
Bytomka	Gliwice	1.83	3.15	2.61	20.6	19.10	151.0

Average values for 1961–1999; maximum values for 1997, with the exception of the Brynica (1983), the Biała Przemsza (1955), the Przemsza (1903) and the Kłodnica (1940)

1974–1984 (Absalon and Wac 1992). In 1994, the share of mine water alone in the actual outflow of major rivers of the study area ranged from 13 % (the Kłodnica) to 80 % (the Biała Przemsza, including nearly half from mine drainage from stowing sands) (Czaja 1999). In the last decade, however, a decreasing trend in the participation of "alien" water in the river outflow has been noticeable (e.g. Jankowski and Szekiel 2009). Table 5.8 summarizes average and maximum water levels and flows in major rivers in the USCB and the corresponding runoff. In addition to the rivers listed in this table, the Rawa, the Pszczynka, and the Bierawka have average flows greater than 3 m^3/s. Discharges of mine water also increased the flows in the Gostynia (1.42 m^3/s), the Mleczna (1.03 m^3/s), the Nacyna (0.94 m^3/s), and the Goławiec Stream (0.84 m^3/s).

Riverbeds of most rivers carrying "alien" water have hydro-technical development, limiting erosion processes. Their intensive course is associated primarily with an extremely high water table due to natural causes. Larger rivers, even if they carry a lot of alien waters, are able to accommodate flood waters if they are embanked or have fortified edges. However, this is not a rule, as shown by the example of flooding in 2010, including the Gostynia valley (Fig. 5.12). A high risk of flooding, especially in areas with small denivelations, may be posed by smaller watercourses that do not have embankments. In the valley of the Stawowy, a tributary of the Mleczna, two subsidence troughs had been formed until 1998, with depths of up to 6.5 m, which covered the area of railway side-track of the Bieruń-Lędziny line and buildings on Hodowlana street in the former Great Bieruń Pond. These areas were flooded during the flood of 2010.

Fig. 5.12 Flood in the area of mining subsidence in the Gostynia valley, Bojszowy Dolne, the Oświęcim Basin (Dulias 2010)

Changes in the stream load. A characteristic feature of many rivers in the study area is above-norm water pollution with suspension (Mackiewicz et al. 1979; Klimek 1993; Szczypek and Wach 1993; Nocoń and Kostecki 2005; Nocoń et al. 2006; Działoszyńska-Wawrzkiewicz 2008). In large rivers, such as the Kłodnica, the Bytomka, the Brynica, the Rawa, and the Przemsza, the largest suspension load was recorded in the years 1955–1975 in connection with the intensive development of coal mining; however, even in the late 1980s, about 250 tonnes of suspension were discharged per day to the rivers of the Upper Silesian Industrial Region (Jankowski 1992). The source of contamination is waste from coal mines (from coal flotation), as well as mine tailings stored on spoil tips in the vicinity of riverbeds and mine tailings embankments. The main ingredient of suspension is coal dust. The increased amount of material carried by rivers accelerates the formation of flood terraces, often overlapping with the mine tailings embankment material accumulated along riverbeds (Wach 1987a, b). A strong dependency between the volume of coal mining and the delivery of coal dust by the Przemsza to young alluvial deposits of the Vistula was pointed out by Klimek (1993) and Czajka (2007). In conditions of high water levels, anthropogenic silts deposited on the bottom and edges of riverbeds are raised again and redeposited, including the floodplains.

Nocoń and Kostecki (2005) reported that probably the highest concentration of suspension was found in the Czarniawka (32,640 mg/dm^3). This river frequently shows a concentration of 2,000–6,000 mg/dm^3. The Bytomka is in the range of 45–177 mg/dm^3 (the average is 87 mg/dm^3) and the Kłodnica is in the range of 3–300 mg/dm^3, with a very high concentration of suspension (up to 1400 mg/dm^3)

occurring below the mouth of the heavily contaminated Czarniawka. Studies by the quoted authors show that within a few years (1998–2004) in all measurement stations along the Kłodnica, the maximum concentration of suspension clearly decreased, such as in Gliwice Sośnica (from 5,562 to 1,433 mg/dm^3) and Gliwice Łabędy (from 4,242 to 131 mg/dm^3). High coal silt pollution characterizes the water of the Nacyna: in the 1970s, its contents in the water reached up to 6,000 mg/dm^3 (Mackiewicz et al. 1979). Wastewater with much suspension silts the riverbeds of the Bierawka, the Bielszowice Stream, the Szotkówka, the Stream from Gogołowa, and the Goławiec Stream. The content of organic carbon in bottom sediments in the estuary section of the Czarniawka reaches 8.5 %; in the Bielszowice Stream, it reaches up to 36.2 % (Działoszyńska-Wawrzkiewicz 2008). A high level of suspension (between 30–50 mg/dm^3) is carried by the Bolina, the Jordanek, and the Bobrek; a little less, but more than 20 mg/dm^3, is carried by the Przyrwa, the Mleczna, and the Michałkowice Ditch.

5.4 Deposition Zone

The matter transported by rivers partly accumulates in riverbeds and in flood conditions outside the riverbeds. The greatest role in "capturing" the material transported by the river is played by water reservoirs (Fig. 5.13). In the nineteenth

Fig. 5.13 The Szotkówka delta at its mouth to the water reservoir in the subsidence trough in Połomia, the Rybnik Plateau (Dulias 2012)

century, fish ponds were of major importance in this respect, located in most valleys in the area of the USCB. Today, water reservoirs of various origins constitute sediment basins for sediment carried by rivers, including the ones that followed the exploitation of sand pits, as well as floodplains in the overdeepening of valleys (subsidence troughs). Insignificant river gradients and their pollution with suspension are conductive to the accumulation of sediments (Dąbrowska et al. 1987; Wach 1987a, b; Eckes and Żuławski 1988).

In the USCB, water reservoirs in subsidence troughs occupy an area of 8–10 km^2. The biggest numbers are in the valleys of the Kłodnica and the Bierawka where their area is, according to various estimates, about 3–4 km^2 (Staszewski et al. 1993). Floodplains are also found in many other valleys of the Oder basin: the Knurówka, the Cienka, the Krywałd, Gierałtowice, and Bielszowice streams, the Nacyna, the Szotkówka, the Jastrzębianka, the Ruptawka, the Marklówka, the Tributary from Michałkowice, and the Vistula basin—the Rawa, the Bolina, the Szarlejka, the Bobrek, the Pszczynka, the Dębinka, the Mleczna, and the Przyrwa (Jankowski and Zobek 1987).

Anthropogenic water reservoirs occurring in the studied mining area were the subject of numerous studies, including studies devoted to the accumulation of bottom sediments (e.g. Rzętała M.A. 2003; Rzętała 2008; Rzętała M.A. et al. 2009; Machowski 2010). Previously, Szczypek and Wach (1993) discussed sedimentation in reservoirs located in river valleys. On the basis of a monthly distribution of the amount of suspension, they found that the Kłodnica, which has been almost entirely transformed by man, carries surprisingly little suspension as the river settles it in the Dzierżno Duże Reservoir above the measurement station of Łany Małe.

Rzętała M.A. (2003) estimated the silting of several water reservoirs created in old sandpits. In the Pogoria III reservoir in the valley of the Pogoria, it amounted to 0.2 % of the capacity of the bowl (524 m^3/year); in the Dzierżno Duże reservoir in the valley of the Kłodnica, it was 2.3 % (55,868 m^3/year). In the first reservoir, bottom sediments have an average thickness of several centimetres; in the other, they are slightly more than 30 cm, but the thickness in the deepest areas reaches several meters. Sediments that build the Kłodnica delta at the mouth to Dzierżno Duże Reservoir consist of gravel fraction of coal dust and sewage sludge. Rzętała (2008) highlighted the rapid pace of sediment delivery to the reservoir, which makes the process of its shallowing very dynamic.

In summary, contemporary conditions of matter circulation in fluvial systems of the Upper Silesian Coal Basin are substantially altered in relation to pre-mining conditions. Data presented in this chapter show that in 90 % of the studied catchment basins, the inclination of slopes increased, 80 % of the catchments increased their relief energy expressed by the relief rate *L*, more than 80 % of rivers lowered their base level of erosion, three-quarters of rivers increased their gradients, and most rivers increased their flows as a result of "alien" water supply. At the same time, almost all the rivers and watercourses underwent at least one type of riverbed geometry change (the straightening, repaving, embankment, paving the bottom, bank reinforcement, the establishment of thresholds, etc.). In many catchments, land use radically changed and anthropogenic forms emerged. These changes disturbed the flow of energy and matter from the pre-mining period. Generalizing, it can be concluded that as a result of mining, the erosion potential of the rivers in the Oder basin was enhanced; in the case of the rivers in the Vistula basin, it weakened. Enhanced removal of the matter from the slopes is mainly implemented in small and medium-sized catchments that are poorly developed and not forested. Some large and medium rivers transport matter over short segments, leaving it in flow-through reservoirs. In many catchments, the slope subsystem does not have any connection with the riverbed, mainly because of the embankments of rivers. Almost 8 % of the mining area was excluded from the fluvial system, resulting in the creation of landlocked basins, mainly in the Triassic zone.

References

Absalon D, Wac M (1992) Antropogeniczne przeobrażenia stosunków wodnych w zlewni Mlecznej. Geogr Stud Dissert 16:9–23

Atlas geologiczno-inżynierski aglomeracji katowickiej 1: 10 000. Katowice, Warszawa, Wrocław, 2005

Brodowski R (2009) Wpływ wilgotności i gęstości gleby lessowej na powierzchniową erozję wodną. Acta Agroph 14(3):567–576

Cabała J, Sutkowska K (2006) Wpływ dawnej eksploatacji i przeróbki rud Zn-Pb na skład mineralny gleb industrialnych, rejon Olkusza i Jaworzna. Prace Nauk Inst Górn Politech Wroc 117. Stud Mat 32:13–22

Cebulak S, Kozłowski K (1978) Charakterystyka mineralogiczno-petrograficzna skał gromadzonych na centralnym zwałowisku „Przezchlebie". Geologia 3:91–99

Czaja S (1988) Zmiany działu wodnego Wisła—Odra w obrębie Górnośląskiego Okręgu Przemysłowego. Geogr Stud Dissert 11:95–100

Czaja S (1992) Zmiany zagospodarowania przestrzennego Górnośląskiego Okręgu Przemysłowego w latach 1860–1985. In Aktualne problemy ekologiczne regionu górnośląskiego 1, Rogoźnik, pp 37–41

Czaja S (1997) Antropogeniczne przeobrażenia powierzchniowej sieci hydrograficznej w zlewni Rawy w latach 1801–1994. Kształ środ geogr ochr przyr obsz uprzem zurb 24:12–18

Czaja S (1999) Zmiany stosunków wodnych w warunkach antropopresji. Wyd. Uniw. Śląskiego, Katowice

Czajka A (2007) Środowisko sedymentacji osadów przykorytowych rzek uregulowanych na przykładzie górnej Odry i górnej Wisły. Prace Nauk UŚ 2534, Katowice

Dąbrowska L, Kłosowski F, Tkocz M, Wrona A (1987) Obszary degradacji środowiska naturalnego w zachodniej części konurbacji górnośląskiej. Mat 36 Zjazdu PTG, Katowice-Sosnowiec, pp 77–87

Dulias R (2006) Possibilities to apply topographic maps on large scale to research on relief changes in mining areas. Anthr aspects transform 4:23–28

Dulias R (2007) Geomorfologiczne skutki eksploatacji węgla kamiennego w Zagłębiu Dąbrowskim. Kształ środ geogr ochr przyr obsz uprzem zurb 38:11–22

Dulias R (2008a) Mining subsidence in Oświęcim Basin (Carpathian Foredeep). Geomorph Slovaca Bohemica 8, 2008/2:7–13

Dulias R (2008b) Wpływ górnictwa węglowego na zmiany krajobrazu w zlewniach Dębinki i Pniówka na Płaskowyżu Rybnickim. Dokum Geogr 37:144–149

Dulias R (2010) Anthropogenic denudation in mining areas: a case study of "Andaluzja" mine. Anthr Aspects Transform 6:23–28

Dulias R (2011) Impact of mining subsidence on the releif of the, Poland. Z Geomorph 55, Suppl 1, Stuttgart, Januar:25–36

Dulias R (2013) Denudacja antropogeniczna na obszarach górniczych na przykładzie Górnośląskiego Zagłębia Węglowego. Wyd Uniw Śląskiego, Katowice

Dwucet K, Krajewski W, Wach J (1992) Rekultywacja i rewaloryzacja środowiska. UŚ, Katowice

Działoszyńska-Wawrzkiewicz M (2008) Metale ciężkie w osadach rzecznych terenów zurbanizowanych zlewni rzeki Kłodnicy. Metale ciężkie w środowisku. Prace Inst Ekol Teren Uprzem

Eckes T, Żuławski CZ (1988) Ochrona terenów górniczych przyrodniczo użytkowanych. Zesz Nauk AGH 1222, Sozologia Sozotechnika 26. Kraków

Greszta J, Morawski S (1972) Rekultywacja nieużytków poprzemysłowych. PWRiL Warszawa

Janecka B, Jabłońska B, Doniecki T, Kacprzak M (2009) Określenie możliwości rekultywacji zwałowiska odpadów cynkowo-ołowiowych przy zastosowaniu kompostu i modyfikacji składu granulometrycznego. In Malina G (ed.) Rekultywacja i rewitalizacja terenów zdegradowanych. Polskie Zrzesz Inż Techn Sanit Poznań, pp 221–238

Jania J (1983) Antropogeniczne zmiany rzeźby terenu wschodniej części Wyżyny Śląskiej. Dokumentacja teledetekcyjna. Prace Nauk Uniw Śląskiego 575:69–91

Jankowski AT (1986) Antropogeniczne zmiany stosunków wodnych na obszarze uprzemysłowionym i zurbanizowanym (na przykładzie ROW). Uniw. Śląski, Katowice

Jankowski AT, Zobek E (1987) Podtopienia terenu na obszarze województwa katowickiego (przyczyny występowania i metody przeciwdziałania). Mat Symp—Problemy geograficzne górnośląsko-ostrawskiego regionu przemysłowego. Katowice–Sosnowiec, pp 42–48

Jankowski AT (1987) Wpływ urbanizacji i uprzemysłowienia na zmianę stosunków wodnych w regionie śląskim w świetle dotychczasowych badań. Geogr Stud Dissert 10:62–99

Jankowski AT (1988) Wpływ przemysłu i urbanizacji na zmiany odpływu Rawy (próba oceny). Dokum Geogr 4:51–63

Jankowski AT (1992) Wpływ gospodarczej działalności człowieka na zmiany warunków hydrologicznych województwa katowickiego. In Aktualne problemy ekologiczne regionu górnośląskiego 1, Rogoźnik, pp 31–36

Jankowski AT, Szekiel M (2009) Ocena zmian warunków hydrologicznych i hydrochemicznych wód rzeki Rawy w pierwszych latach XXI wieku. Kształ środ geogr ochr przyr obsz uprzem zurb 40:75–89

Kaszowska O (2005) Koszty usuwania szkód górniczych w kopalniach Górnośląskiego Zagłębia Węglowego. In: Kwiatek J (ed) Problemy eksploatacji górniczej pod terenami zagospodarowanymi. Ustroń, pp 242–251

Klimek K (1993) Środowisko sedymentacji antropogennych osadów pozakorytowych w dolinach Przemszy i Wisły Śląskiej. In: Klimek K (ed) Antropogenne aluwia Przemszy i Wisły Śląskiej. Georama 1. UŚ Sosnowiec, pp 3–15

Konstantynowicz E (1989) Wpływ rozwoju bazy surowcowej województwa katowickiego na stan środowiska przyrodniczego. Mat Konf - Bariery funkcjonowania Górnego Śląska, Katowice, pp 65–90

Kupka R, Frolik H, Dulias R (2008) Zmiany rzeźby na obszarze górniczym zlikwidowanej kopalni „Katowice-Kleofas". Informacja ogólna. Kształ środ geogr ochr przyr obsz uprzem zurb 39:26–31

Kupka R, Szczypek T, Wach J (2005) Morphological effect of 200-years long hard coal exploitation in Katowice. In: Szabó J, Morkūnaitė R (eds.) Landscapes—nature and man. Univ Debrecen, Lithuanian Inst Geol Geogr, Debrecen-Vilnius, pp 95–100

Machowski R (2010) Przemiany geosystemów zbiorników wodnych powstałych w nieckach osiadania na Wyżynie Katowickiej. Wyd Uniw Śląskiego, Katowice

Maciak F (2003) Ochrona i rekultywacja środowiska. Wyd SGGW, Warszawa

Mackiewicz D, Miksa K, Mrozowska A (1979) Program uregulowania stosunków wodnych na obszarach objętych szkodami górniczymi w województwie katowickim. CBSiPBW Hydroprojekt, Katowice

Mapa topograficzna 1: 10 000. Główny Geodeta Kraju, 1993–1994
Mazurek M, Zwoliński Z (2012) System rzeczny (fluwialny). http://hum.amu.edu.pl/sgp/gw/sf/sf.html. Inst Bad Czwart UAM, Poznań [2012.03.21]
Migoń P (2006) Geomorfologia. Wyd Nauk PWN, Warszawa
Molenda T (1999) Wpływ działalności górniczej na kształtowanie stosunków wodnych (na wybranych przykładach z Górnośląskiego Zagłębia Węglowego. In: Pełka-Gościniak J, Rzętała M (eds) Górnośląsko-Ostrawski region przemysłowy—wybrane problemy ochrony i kształtowania środowiska. Sosnowiec, pp 154–158
Nocoń W, Kostecki M (2005) Hydro-chemical characteristic of the Czarniawka River. Arch Ochr Środ 2:95–104
Nocoń W, Kostecki M, Kozłowski J (2006) Charakterystyka hydrochemiczna rzeki Kłodnicy. Ochr Środ 28(3):39–44
Ostrowski J (ed) (2001) Ochrona środowiska na terenach górniczych. Wyd. CCCPSMiE PAN, Kraków
Posyłek E, Rogoż M (1982) Wpływ osiadania powierzchni terenu na stosunki wodne na obszarze miasta Katowice. Mat Konf—Ochrona środowiska na terenie aglomeracji Katowice. Warszawa-Sosnowiec, pp 16–26
Pozzi M, Cempiel E, Czajkowska A (2008) Koncepcja regulacji stosunków wodnych na terenie górniczym gminy Gierałtowice. Gosp Sur Min 24, 2/3:109–122
Rosik-Dulewska Cz (2006) Podstawy gospodarki odpadami. Wyd. Nauk PWN, Warszawa
Rzętała M (2008) Funkcjonowanie zbiorników oraz przebieg procesów limnicznych w warunkach zróżnicowanej antropopresji na przykładzie regionu górnośląskiego. Wyd Uniw Śląskiego, Katowice
Rzętała MA (2003) Procesy brzegowe i osady denne wybranych zbiorników wodnych w warunkach zróżnicowanej antropopresji (na przykładzie Wyżyny Śląskiej i jej obrzeży). Wyd Uniw Śląskiego, Katowice
Rzętała MA, Machowski R, Rzętała M (2009) Sedymentacja w strefie kontaktu wód rzecznych i jeziornych na przykadzie zbiorników wodnych regionu górnośląskiego. UŚ WNoZ, Sosnowiec
Schumm SA (1977) The fluvial system. Wiley, New York
Staszewski B, Augustyniak I, Bukowski P, Górka G, Strzemiński T (1993) Ocena zasobów wodnych zalewisk i możliwości ich zagospodarowania w wybranym rejonie GZW z użyciem komputerowego systemu symulacji. Dokum nr 2216073BH. Arch. GIG. Katowice
Strzelczuk H (1977) Wybrane zagadnienia gospodarki odpadami. IKŚ, Warszawa
Szczepańska J (1987) Zwałowiska odpadów węgla kamiennego jako ogniska zanieczyszczeń środowiska wodnego. Zesz Nauk AGH, Geologia 35
Szczypek T, Wach J (1991a) Rozwój współczesnej wydmy w warunkach silnej antropopresji. Wyd Uniw Śląskiego, Katowice
Szczypek T, Wach J (1991b) Human impact and intensity of in the Silesian-Cracow Upland (Southern Poland). Z. Geomorph. N.E, Suppl.-Bd 90, Berlin-Stuttgart:171–177
Szczypek T, Wach J (1993) Miesięczny rozkład ilości zawiesiny w wybranych rzekach regionu śląsko-krakowskiego. In: Křiž V, Prášek J, Jankowski A (eds.) Změny geografickěho prostředí v pohraničních oblastech Ostravského a Hornoslezského regionu. Ostrava, pp 135–141
Wach J (1987a) Zmiany profilu podłużnego Kłodnicy w wyniku osiadań górniczych. In: Mat Symp—Problemy geograficzne górnośląsko-ostrawskiego regionu przemysłowego. Katowice-Sosnowiec, pp 126–130
Wach J (1987b) Antropogeniczne formy rzeźby—zagadnienie nomenklatury. Mat. 36 Zjazdu PTG. Katowice- Sosnowiec, pp 16–17
Wach J (1991) Wpływ antropopresji na kształtowanie się rzeźby terenu. Mat. Symp—Człowiek i jego środowisko w górnośląsko-ostrawskim regionie przemysłowym. UŚ WNoZ Sosnowiec, pp 115–119

Wach J, Szczypek T (1996) Preobrazovania rel`efa mestnosti v raionakh gornodobyvaiushchei promyshlennosti vsledstvie osedanii grunta (na primere Katovickovo voevodstva). In: Pirozhnik II (ed) Geograficheskie problemy prirodopolzovania v usloviakh antropogennoi deiatelnosti. Belorusskii Gosudarstvennyi univ., Belorusskoe Geograficheskoe obshchestvo, Minsk, pp 21–27

Wilk Z (ed) (2003) Hydrogeologia polskich złóż kopalin i problemy wodne górnictwa. Uczel Wyd Nauk Dydakt, Kraków

Wrona A (1975) Problemy degradacji i ochrony powierzchni ziemi w Rybnickim Okręgu Węglowym. Przegl Geogr 47(3):519–538

Chapter 6
Changes in the Circulation of Matter in Landlocked Basins

Landlocked basins constitute a characteristic feature of the relief in central Europe, especially in post-glacial areas. In Poland, they have been studied mainly in hydrological and geomorphological aspects (e.g. Werner-Więckowska 1953; Maruszczak 1954; Klajnert 1965; Kowalska 1968; Drwal 1975; Wieczorkowska 1976; Borówka 1992). There is a steady inflow of matter from the atmosphere to the landlocked basins, and its circulation takes place due to the water cycle in their catchment areas. The underground runoff may be several tens of percent of the atmospheric supply (Drwal 1982); therefore, catchments of such landlocked basins may be treated as open systems, despite the lack of drainage to the surface (Major 2009). The catchments wholly without drainage constitute closed systems, which result from the lack of permeability of surface deposits. The matter is transported from the watershed on the slope to the bottom of the basin, where sedimentation takes place. The circulation of matter is strongly dependent on precipitation and relative heights, which reflect the potential energy (Kostrzewski 1986).

In the USCB in the pre-mining period, there were only a few landlocked basins, which were wind-blown troughs or depressions between dunes, oxbow lakes, or concaves on high plains, among others. Today, these forms are characteristic features of the relief. The genesis of the basins is mostly anthropogenic, resulting primarily from conducted mining operations. Subsidence troughs constitute the dominant group of landlocked basins (Wach and Szczypek 1996; Perski 2000; Madowicz 2001; Dulias 2005, 2008, 2010; Dulias and Szczypek 2005; Kupka et al. 2005; Wojciechowski 2007; Solarski and Pradela 2010a). A broader characteristic of landlocked forms was provided by Dulias (2003a, b), who presented their morphometric features: sizes, shapes, inclinations, and lengths of slopes, as well as observations on contemporary geomorphological processes using examples from the area of Piekary Śląskie on the Bytom Plateau.

Landlocked basins examined in this study were classified as closed systems only on the basis of the lack of surface drainage. Their range was determined based on topographic maps of 1: 10,000, depicting the relief in 1993. All of the depressions from which there is no drainage of surface water (fluvial, slopewash, anthropogenic

R. Dulias, *The Impact of Mining on the Landscape*,
Environmental Science and Engineering, DOI 10.1007/978-3-319-29541-1_6

Fig. 6.1 Water reservoir in the subsidence trough of the Szczygłowice mine on the Wilcza High Plain (Dulias 2006)

drainage) and which are limited by natural or anthropogenic watersheds were considered to be land-locked. Some landlocked basins were formed by dividing a valley by a spoil tip (Fig. 6.1); several forms appeared due to their "fencing off" from the river by flood embankments. Small concaves within the bottoms of landlocked basins were not isolated as separate forms; instead, the principle of inclusion of distinct but small landlocked concaves into the adjacent larger concaves was adapted, for example, in areas of crossings of high railway embankments. The analysis includes concaves of areas larger than 0.01 km^2, which is why it does not include minor forms of open-pit mining.

A total of 233 land-locked basins were identified, with a total area of 122.3 km^2 (4.3 % of the area of research and 7.6 % of the mining area) (Fig. 6.2). This is a size 4 times larger than that predicted by Mackiewicz et al. (1979) for the province of Katowice for 1990 (31 km^2). About three-quarters of the forms occur in the north central part of the USCB in two continuous strips, one of which is of almost the same shape as the occurrence of the Saddle Beds. The lithology of cover deposits in the occurrence of landlocked basins is not very diverse due to the presence of mostly clay deposits on the surface.

Landlocked basins are primarily concentrated in the Triassic zone, where more than 56 % of all forms (131) are found with a total area of 74.6 km^2, which makes up 61 % of the total area of landlocked basins. In the Carboniferous zone, 50 forms were identified, covering an area of 23 km^2 (18.8 %). In the Miocene zone are 52

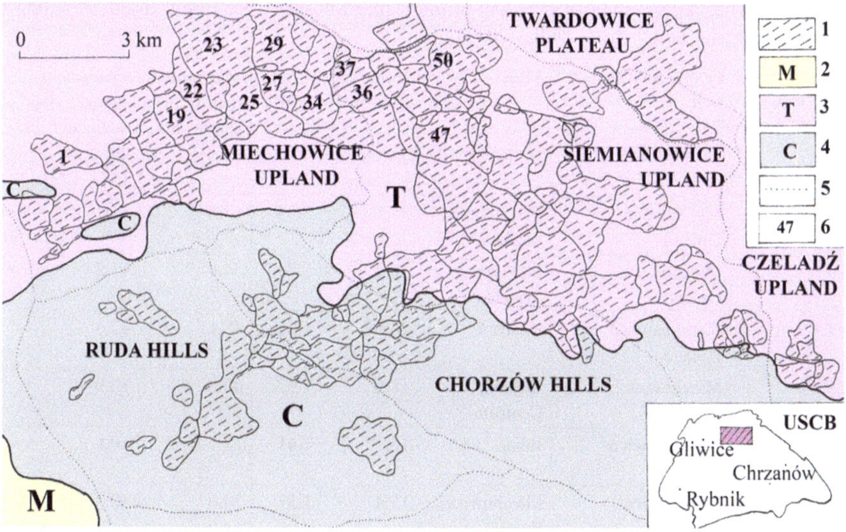

Fig. 6.2 Location of land-locked basins in the northern part of the Upper Silesian Coal Basin in 1993 (from Dulias 2013). (**1**) Land-locked basins, (**2**) Miocene zone, (**3**) Triassic zone, (**4**) Carboniferous zone, (**5**) boundaries of geomorphological units, (**6**) numbers of closed basins listed in Table 6.1

forms with a total area of 24.7 km^2 (20.2 %). The dominant basins are dry, with 82 hydrated forms is 82 (35.5 %). Half of the depressions are in the Carboniferous zone, 36 % in the Triassic, and 21 % in the Miocene zone.

Landlocked basins occur in 19 geomorphological units, but they constitute important elements of the relief in six of them. An absolutely unique position in this regard is held by the Siemianowice Plateau, where landlocked forms occupy almost 40 km^2, which means that as much as half of the area of this geomorphological unit was excluded from fluvial drainage. Landlocked basins occupy a total of 23.7 km^2 in the Miechowice Upland (22 % of the Upland), 15.8 km^2 (3.6 %) on the Rybnik Plateau, 11.1 km^2 (24.2 %) on the Chorzów Hills, 8.4 km^2 (15.5 %) on the Ruda Hills, and 7.8 km^2 (3.2 %) in the Rachowice Upland.

The described forms are diverse in terms of size. In the Triassic zone, landlocked basins cover an area in the range of 0.03–3.2 km^2, averaging 0.57 km^2, wherein as many as 24 forms have a surface area greater than 1 km^2. In the other two zones—the Carboniferous and the Miocene—the surface covered by landlocked basins is smaller; on average, it is 0.46 km^2, ranging from 0.03 to 2 km^2. In each of the zones, there are 6 forms with an area exceeding 1 km^2. The widths of the forms are varied, but they usually amount to about 1 km (±200–300 m). Table 6.1 summarizes basic data for the largest landlocked basins in the USCB area.

Landlocked basins limited by natural watersheds are oval or elliptical forms. The shapes of the remaining forms—with partially artificial boundaries extending on culminations of spoil tips and landfills, as well as rail and road embankments—are

Table 6.1 The largest landlocked basins in the Upper Silesian Coal Basin, listed according to geomorphological units[a] (from Dulias 2013)

Number of land-locked basin	Geomorphological unit	Mine	Volume (mln m^3)	Area (km^2)	Maximum relative height (m)	Changes in average slope inclination in the years 1883–1993 (°)
23	Miechowice Upland	Powstańców Śląskich	34.5	3.23	17.5	+1.1
29	Miechowice Upland	Powstańców Śląskich	30.9	1.74	34.4	+1.2
47	Siemianowice Upland	Rozbark	26.0	1.66	48.1	+2.3
25	Miechowice Upland	Bobrek, Centrum	18.5	1.97	40.0	+0.9
50	Siemianowice Upland	Julian	16.7	1.41	28.7	+0.8
19	Miechowice Upland	Miechowice, Bobrek	15.4	1.37	30.0	+0.3
27	Miechowice Upland	Centrum	14.2	0.77	27.8	+1.6
1	Miechowice Upland	Pstrowski	13.5	1.10	18.3	+2.2
34	Miechowice Upland	Centrum	13.4	1.06	22.5	+0.7
37	Siemianowice Upland	Centrum, Powstańców Śląskich	11.2	0.74	27.4	+1.1
22	Miechowice Upland	Bobrek	10.5	1.03	24.5	+0.4
36	Siemianowice Upland	Centrum	–	0.93	33.7	+1.6

[a]The numbering of forms is consistent with the numbering in Fig. 6.2

more varied. Many troughs are divided by embankments into several smaller forms with irregular shapes.

Together with the separation of landlocked basins from river catchments, the conditions of the circulation of matter changed. The movement of surficial sediments is directed to the inside of forms; this is why, from the geomorphological point of view, post-mining concave forms make the new land-locked sedimentary basins. Wach (1987, 1991) expressed the opinion that the occurrence of a large number of small land-locked basins makes the USCB area similar to a young glacial area, especially in reference to the large amount of water reservoirs. In the geographical literature, the area is referred to as the Upper Silesian anthropogenic lake district (Jankowski 1986; Czaja 2003; Rzętała 2008). The location of landlocked basins presented in Fig. 6.2 and their physiognomy in Fig. 6.3 also impose the similarity of the landscape (not genetic) to a young glacial area.

Fig. 6.3 Small landlocked basin in the subsided area of Bytom Rozbark in the Siemianowice Upland (Dulias 2003)

The movement of matter in landlocked basins was changed primarily in terms of the direction and length of the transport of matter due to slope washing. The reversal of the natural gradient of the terrain led to the direction of the surface flow to the centre of the trough, with a clear shortening of its way to the base level of erosion (Fig. 6.4). In most basins, the dominating slopes are 300–700 m long; only in large forms are they longer than 1 km. Denivelation within landlocked basins has a very wide range—from 3.2 to 117.5 m, with an average of 20–22 m. Relative heights over 40 m are usually associated with the occurrence of a spoil tip in the watershed area of the basin; such relative heights were observed in 15 forms. In 84 % of the forms, their separation in the relief was associated with increased slope inclination, usually by 1°; however, in 74 basins (32 %), an average gradient increased by more than 1°, up to 4.3°. The inclination of slopes increased most on the Rybnik Plateau, by an average of 1.5°. On this loess area, intensification of the slopewash process is to be expected.

In landlocked basins with water reservoirs, the participant of the circulation of matter, apart from slopewashing, is the water wave. The geographical literature contains very little information on shoreline processes in this type of water reservoir. Dulias and Rudnicka (2000) presented the classification of shores on the basis of research of reservoirs in the forks of the Brynica and the Rawa. Rzętała (2003), when examining shoreline processes in water reservoirs, also made comments about their progress in certain landlocked basins. However, Machowski (2010) presented the broadest view on the subject by characterizing water reservoirs in subsidence troughs in the Katowice Upland. These authors emphasized that shoreline processes depend on wind waves to the greatest extent. Machowski (2010) observed ephemeral microcliffs in the area of Żabie Doły on the Bytom Plateau, with heights up to 20 cm, and the more durable microbays as well as microterraces in one of the

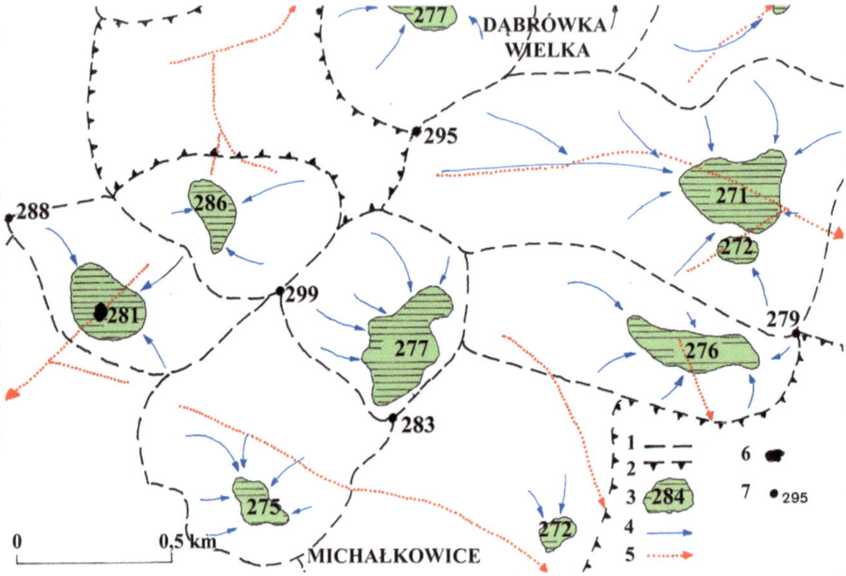

Fig. 6.4 Changes in the directions of matter movement in the Siemianowice Upland due to the formation of landlocked basins (based on Dulias 2003b). (**1**) Natural watershed, (**2**) anthropogenic watershed, (**3**) bottoms of closed basins with marked absolute height, (**4**) contemporary directions of matter movements, (**5**) main directions of matter movements in the pre-mining period, (**6**) water reservoirs, (**7**) elevation points in meters above sea level

reservoirs in Zabrze Makoszowy. The author pointed out that the activities of waves are greater during the almost-annual summer lowering of the water level in the reservoirs. The microterraces in the deepest landlocked subsidence trough in the USCB were described by Solarski and Pradela (2010b), who associated them with the gradual lowering of the water level in the basin from 2005 onwards. Some importance in shaping the shores is also given to the ice phenomena, mainly in the period of disappearance of ice floes.

The circulation of matter, both in open systems as well as in landlocked basins, is also implemented by aeolian manner. However, in landlocked basins, due to their small sizes, it seems to play a greater role in the balance of transferred matter. During field research, the process of both anthropogenic and natural sediment dissemination was observed many times. The agricultural land of the Siemianowice Upland is particularly prone to deflation, as well as the mine tailing heaps on the Rybnik Plateau, where clouds of dust hovering over spoil tips devoid of vegetation were sometimes visible from a distance of over 2 km. Favourable conditions for the occurrence of aeolian processes are created by lithological features of the surface sediments of the study area—dust and clay deposits and fine-grained material stored on the tips, especially in settling tanks. Aeolian processes also take place in sandpits; however, most of these forms, due to the system of drainage ditches, do not fall into the category of landlocked basins. In landlocked forms, regardless of the winnowing of desiccated autochthonous sediments, deposition of allochthonous sediments takes place. The nature of changes in the circulation of matter through the atmosphere is reflected in changes in atmospheric pollution in the last half-century, expressed by dust fallout: in the 1970s, mostly exceeding 1,000 tonnes/km^2 (maximum of 2,000 tonnes/km^2), and in the late 1990s reaching tens of tonnes per square kilometre of the surface (maximum of 200 tonnes/km^2) (Leśniok and Degórska 2008).

The essence of the circulation of matter in land-locked basins, both dry and watered, is its accumulation on the bottom of the forms. This process involves the mineral matter from slope wash, the blurring of shores, from dry and wet deposition from the atmosphere and decomposed organic matter. According to data compiled by Rzętała (2003), the thickness of bottom sediments in water reservoirs in subsidence troughs ranges from a few to 25–30 cm.

To sum up, the formation of landlocked basins has changed the conditions of the circulation of matter, especially in the area of the Vistula–Oder watershed, on an area built from Triassic rocks with a cover of Quaternary deposits of clay. Mostly the upper segments of valleys were excluded from the fluvial system. Most basins have become apparent in the relief over the past several dozen years, and the upper layers of sediments accumulated in these new anthropogenic sedimentary basins are young. The potential duration of a lack of drainage depends on many factors, including targeted measures aimed at its interruption and morphometric features of forms. The functioning period of landlocked basins is dependent on the minimum relative height in the place of the possible reinclusion of its territory in the area of the river catchment; in some (but very few) forms, it amounts to only 1 m.

References

Borówka RK (1992) Przebieg i rozmiary denudacji w obrębie śródwysoczyznowych basenów sedymentacyjnych podczas późnego vistulianu i holocenu. UAM Ser Geogr 54, Poznań

Czaja S (2003) Zbiorniki i pojezierza antropogeniczne. In: Szczypek T, Rzętała M (eds) Człowiek i woda. PTG Oddz Katowicki, Sosnowiec, pp 22–30

Drwal J (1975) Zagadnienia bezodpływowości na obszarach młodoglacjalnych. Zesz Nauk WBiNoZ Uniw Gdański, Geografia 3:7–26

Drwal J (1982) Wykształcenie i organizacja sieci hydrograficznej jako podstawa oceny struktury odpływu na terenach młodoglacjalnych. Zesz Nauk, Rozprawy Monografie 33, Uniw Gdański, Gdańsk

Dulias R (2003a) Bezodpływowe baseny sedymentacyjne na obszarze osiadań górniczych na Płaskowyżu Bytomskim. In: Rzętała M, Jankowski AT (eds) Problemy geoekologiczne górnośląsko-ostrawskiego regionu przemysłowego, Sosnowiec, pp 23–27

Dulias R (2003b) Subsidence depressions in Upper Silesian Coal Basin. In: Mentlik P (ed) Geomorfologický sborník 2:11–16

Dulias R (2005) Krzywe hipsograficzne obszaru osiadań górniczych (na przykładzie okolic Piekar Śląskich). In: Kotarba A, Krzemień K, Święchowicz J (eds) Współczesna ewolucja rzeźby Polski. Kraków: 115–120

Dulias R (2008) Wpływ górnictwa węglowego na zmiany krajobrazu w zlewniach Dębinki i Pniówka na Płaskowyżu Rybnickim. Dokum Geogr 37:144–149

Dulias R (2010) Anthropogenic denudation in mining areas: a case study of "Andaluzja" mine, Silesian Upland. Anthropogenic aspects landscape Transformations 6:23–28

Dulias R (2013) Denudacja antropogeniczna na obszarach górniczych na przykładzie Górnośląskiego Zagłębia Węglowego. Wyd Uniw Śląskiego, Katowice

Dulias R, Rudnicka M (2000) Typy brzegów antropogenicznych zbiorników wodnych na obszarze między Sosnowcem, Katowicami i Mysłowicami. Kształt środ geogr ochr przyr obsz uprzem zurb 30:7–14

Dulias R, Szczypek T (2005) Badania antropogenicznych basenów sedymentacyjnych Wyżyny Śląskiej z zastosowaniem ^{137}Cs. Geomorfologický sborník 4:15–18

Jankowski AT (1986) Antropogeniczne zmiany stosunków wodnych na obszarze uprzemysłowionym i zurbanizowanym (na przykładzie ROW). Uniw Śląski, Katowice

Klajnert Z (1965) Budowa geologiczna i geneza zagłębień bezodpływowych w Józefowie. Przegl Geogr 37:143–162

Kostrzewski A (1986) Zastosowanie teorii funkcjonowania geoekosystemu do badań współczesnych środowisk morfogenetycznych obszarów nizinnych Polski Północno-Zachodniej. Spraw PTPN 103:26–28

Kowalska A (1968) Obszary bezodpływowe środkowej części Niżu Polskiego. Wyd UMCS, WBiNoZ, Lublin

Kupka R, Szczypek T, Wach J (2005) Morphological effect of 200-years long hard coal exploitation in Katowice. In: Szabó J, Morkūnaitė R (eds) Landscapes – nature and man. Univ Debrecen, Lithuanian Inst. Geol Geogr, Debrecen-Vilnius: 95–100

Leśniok M, Degórska V (2008) Zanieczyszczenie powietrza w miastach Górnośląskiego Związku Metropolitalnego. In: Dulias R, Hibszer A (eds) Górnośląski Związek Metropolitalny. Zarys geograficzny, PTG Oddz Katowicki, Sosnowiec, pp 105–118

Machowski R (2010) Przemiany geosystemów zbiorników wodnych powstałych w nieckach osiadania na Wyżynie Katowickiej. Wyd Uniw Śląskiego, Katowice

Mackiewicz D, Miksa K, Mrozowska A (1979) Program uregulowania stosunków wodnych na obszarach objętych szkodami górniczymi w województwie katowickim. CBSiPBW Hydroprojekt, Katowice

Madowicz A (2001) Osiadania terenu na obszarze Jastrzębia Zdroju w latach 1974–1997. Kształt środ geogr ochr przyr obsz uprzem zurb 31:15–21

Major M (2009) Charakter i funkcjonowanie zagłębień bezodpływowych w krajobrazie strefy młodoglacjlanej (Pomorze Zachodnie, górna Parsęta). PTPN, Prace Kom Geogr Geol 40, Poznań

Maruszczak H (1954) O oczkach lodowcowych i zagłębieniach bezodpływowych. Czas Geogr 25:1–2

Perski Z (2000) Zastosowanie satelitarnej interferometrii radarowej do określania dynamiki i zasięgu górniczych deformacji terenu na przykładzie wybranych obszarów Górnośląskiego Zagłębia Węglowego. Prace WNoZ UŚ 8, Sosnowiec; 9–39

Rzętała MA (2003) Procesy brzegowe i osady denne wybranych zbiorników wodnych w warunkach zróżnicowanej antropopresji (na przykładzie Wyżyny Śląskiej i jej obrzeży). Wyd Uniw Śląskiego Katowice

Rzętała M (2008) Funkcjonowanie zbiorników oraz przebieg procesów limnicznych w warunkach zróżnicowanej antropopresji na przykładzie regionu górnośląskiego. Wyd Uniw Śląskiego, Katowice

Solarski M, Pradela A (2010a) Przemiany wybrany form rzeźby Wyżyny Miechowickiej w latach 1883–1994. Z bad wpł antrop środ 11:78–92

Solarski M, Pradela A (2010b) Przebieg zjawisk lodowych w zbiorniku wodnym w niecce osiadania w sezonie zimowym 2008/2009. Kształt środ geogr ochr przyr obsz uprzem zurb 42:70–79

Wach J (1987) Antropogeniczne formy rzeźby – zagadnienie nomenklatury. Mat. 36 Zjazdu PTG, Katowice, Sosnowiec: 16–17

Wach J (1991) Wpływ antropopresji na kształtowanie się rzeźby terenu. Mat. Symp – Człowiek i jego środowisko w górnośląsko-ostrawskim regionie przemysłowym. UŚ WNoZ, Sosnowiec: 115–119

Wach J, Szczypek T (1996) Preobrazovania rel`efa mestnosti v raionakh gornodobyvaiushchei promyshlennosti vsledstvie osedanii grunta (na primere Katovickovo voevodstva). In: Pirozhnik II (ed) Geograficheskie problemy prirodopolzovania v usloviakh antropogennoi deiatelnosti. Belorusskii Gosudarstvennyi univ, Belorusskoe Geograficheskoe obshchestvo, Minsk, pp 21–27

Werner-Więckowska H (1953) Obszary bezodpływowe Mazowsza. Przegl Geogr 23

Wieczorkowska J (1976) Rola zagłębień bezodpływowych w rozwoju rzeźby okolic Łodzi. Acta Geogr Lodz 37:183–189

Wojciechowski T (2007) Osiadanie powierzchni terenu pod wpływem eksploatacji węgla kamiennego na przykładzie rejonu miasta Knurowa. Przegl Geol 55(7):589–594

Chapter 7
Anthropogenic Denudation Rate in the Upper Silesian Coal Basin

Anthropogenic denudation is mainly related to agriculture and mining activities (e.g. Zapletal 1969; Słupik 1973; Jania 1983; Sinkiewicz 1998; Price et al. 2011; Bennett et al. 2015). Most of the works focus on anthropogenic mechanical denudation, which is understood as the movement of rock matter caused directly and/or indirectly by the economic activity of humans. The concept of anthropogenic denudation is inseparable from anthropogenic aggradation (Zapletal 1968; Demek 1973; Kozarski and Rotnicki 1978; Dulias 2013). Quantification of the intensity of both processes is one of the most important tasks of modern geomorphology.

The history of the economic development of the Upper Silesian Coal Basin indicates that anthropogenic denudation here is largely caused by mining. The first estimate of the scale of the process for the area of the Upper Silesian conurbation was presented by Żmuda (1973), who stated that the volume of subsidence by the end of the 1960s had been 0.9 billion m^3, which created an average lowering of the surface by 1.4 m. Applying this value to about 70 years (at the time) of intense mining activity, an average rate of surface lowering is calculated at approximately 20 mm/year. The quoted author also estimated the volume of ground masses transported during various construction works, opencast mining of raw materials, waste disposal, and land levelling at more than 781 million m^3, which, when evenly distributed within the conurbation, would give a layer with a thickness of about 97 cm.

Research on anthropogenic denudation in the eastern part of the Silesian Upland reveals that in the catchment of the Biała Przemsza, which is situated within intensive exploitation of stowing sands, the mechanical anthropogenic denudation rate in the 1970s was 17 mm/year (Aparta and Jania 1980; Jania 1983), which was almost 600 times greater than the rate of natural global denudation estimated for this part of the Upland at 0.03 mm/year (Dębski 1959). The volume of anthropogenic denudation in the area of coal mining is most often estimated by various authors at 20–60 mm/year (Madowicz 2001; Dulias 2005, 2007a, b, 2010, 2011; Dulias and Szczypek 2005; Kupka et al. 2005, 2008; Wojciechowski 2007; Aleshina et al. 2008; Łajczak 2009); however, in some areas, on short-term scales, it might be as

R. Dulias, *The Impact of Mining on the Landscape*,
Environmental Science and Engineering, DOI 10.1007/978-3-319-29541-1_7

high as several hundred millimeters per year (Perski 2000; Mirek and Isakow 2009).

In this work, the term *anthropogenic denudation* (D_A) is used to describe mechanical denudation resulting from human activities—direct (D_{AD}) or indirect (D_{AI}); a similar distinction was made between anthropogenic aggradation (A_A), with direct (A_{AD}) and indirect (A_{AI}). This chapter presents anthropogenic denudation rates for the area of the USCB, calculated by two methods. The first one refers to the method proposed by Żmuda (1973) and partly by Jania (1983); it is based on statistics of raw material and waste rock output and the corresponding operation coefficients. The other method is based on a morphometric analysis using digital relief models for 1883 and 1993. The measure of anthropogenic denudation calculated by each method is the rate of surface lowering in millimetres per year.

7.1 Anthropogenic Denudation Rates Calculated from Raw Materials Output

In the Upper Silesian Coal Basin, approximately 13.24 billion tonnes of minerals were extracted by 2009, of which as much as 93.8 % is attributable to coal and stowing sands, and only 6.2 % to other raw materials such as zinc and lead ores, iron ores, aggregates, dolomites, limestones, marls, dimension and crushed stones, and clays (Table 7.1). Loss of rock mass from the bedrock of the USCB is therefore almost entirely associated with the mining of coal and stowing sands. Taking into account the further extraction of waste rock in coal mines, estimated at 2.1–4.3 billion tonnes, and the fact that almost 90 % of coal extraction took place in the twentieth century, the mining of coal and stowing sands resulted in the removal of 20 times more rock material in 100 years than mining all other raw materials put together in the last 1,000 years. Quarries, clay pits, *warpia*, sinkholes, and flotation tanks—all anthropogenic forms associated with the extraction of zinc, lead, iron ores and other, excluding stowing sands, despite the apparent presence in the relief of the USCB—account for only 5–6 % of the volume of subsidence troughs and

Table 7.1 Raw materials output in the Upper Silesian Coal Basin until 2009 (from Dulias 2013)

Raw material	Mining area (km^2)	Output to 2009 (mln tonnes)	Percentage share of total output in USCB
Coal	1777.6	10,667	80.55
Stowing sands	76.3	1,755	13.25
Gravel aggregates	15.0	300	2.27
Solid rocks	6.1	280	2.12
Zinc and lead ores	160.0	200	1.51
Iron ores	7.0	20	0.15
Clayey raw materials	3.0	20	0.15

Table 7.2 Anthropogenic denudation rates calculated from the coal and waste rocks output in the Upper Silesian Coal Basin for the years 1769–2009[a] (after Dulias 2013)

Characteristic	Silesian Upland	Racibórz-Oświęcim Basin	Mining area
Mining area (km^2)	921.8	855.8	1777.6
Coal output until 2009 (mln tonnes)	7342.9	3123.5	10 466.4
To 1882	118.2	1.9	120.1
1883–1993	6453.8	2278.6	8732.4
1994–2009	770.9	843.0	1613.9
Subsidence volume to 2009 (mln m^3)	3379.6	1631.9	5011.5
To 1882	67.5	1.1	68.6
1883–1993	2921.6	1155.0	4076.6
1994–2009	390.5	475.8	866.3
Average surface lowering to 2009 (m)	3.7	1.9	2.8
To 1882[b]	0.17–0.34	0.04	0.17–0.34
1883–1993	3.2	1.3	2.3
1994–2009	0.4	0.6	0.5
Anthropogenic denudation rate to 2009 (mm/year)	15	11	12
To 1882[b]	1–3	1	1–3
1883–1993	29	12	21
1994–2009	25	37	31

[a]Excluding the small mines
[b]Calculation for the then mining area

sandpits. Changes in the relief related to the mining of coal are regional and cover a large surface, whereas the mining of other raw materials has a local impact. A further analysis therefore concerns only the anthropogenic denudation related to the mining of coal and stowing sands.

The volume of subsidence in the area of coal mining in the USCB, calculated on the basis of the balance of output according to the method described in Sect. 1.2, accounts for more than 5 billion m^3, which gives an average surface lowering of 2.8 m (Table 7.2). Mining areas located in the Racibórz-Oświęcim Basin decreased by an average of 1.9 m, while those located within the Silesian Upland decreased on average by twice as much, at 3.7 m. The average surface lowering of the area under the influence of mining was 4.2 m. Individual mines differ in terms of an average surface lowering. Until 2009, the biggest decrease was calculated in the following mines: Centrum, 17.2 m; Szombierki, 10.8 m; and Wawel, 10.4 m. The smallest decreases occurred in the youngest mines of Morcinek (0.3 m) and Żory (0.4 m), as well as in the large mine of Gliwice (0.4 m). Table 7.3 lists the volume of anthropogenic denudation in relation to the total area of particular coal mines, and Fig. 7.1 shows the average size of the D_A with respect to the area of mining impact.

Table 7.3 Anthropogenic denudation rates in the areas of coal mines in the Upper Silesian Coal Basin, calculated from the coal and waste rocks output (from Dulias 2013)

No.	Mine	Subsidence volume to 2009 (mln m³)[a]	Average surface lowering in the years 1883–1993 (m)	Anthropogenic denudation rate (mm/year)[b]		
				To 1882	1883–1993	1994–2009
Silesian Upland						
1.	Andaluzja	62.9	6.5	–	79	83
2.	Barbara-Chorzów	62.3	4.1	16	37	–
3.	Bielszowice	251.5	6.2	4	56	49
4.	Bobrek	88.5	6.4	–	76	82
5.	Bolesław Śmiały	89.1	1.0	–	9	13
6.	Centrum	100.7	14.7	–	159	154
7.	Czeladź-Milowice	47.9	7.6	2	68	18
8.	Grodziec	26.1	1.1	–	11	20
9.	Halemba	81.0	2.3	–	58	84
10.	Jan Kanty	31.4	1.2	–	16	11
11.	Jaworzno	129.2	2.3	1	21	31
12.	Jowisz	25.9	2.1	–	26	32
13.	Julian	39.3	2.9	–	73	59
14.	Katowice	49.6	5.5	7	50	17
15.	Kazimierz-Juliusz	48.6	1.7	–	17	18
16.	Kleofas	75.6	4.4	–	41	26
17.	Miechowice	62.3	4.9	–	53	42
18.	Murcki	82.4	1.3	–	11	30
19.	Mysłowice	58.7	4.1	1	37	54
20.	Niwka-Modrzejów	31.2	1.4	–	14	17
21.	Nowy Wirek	63.4	7.5	25	68	96
22.	Paryż	50.7	1.7	1	16	4
23.	Pokój	108.3	8.3	3	74	102
24.	Polska	171.1	6.1	6	55	38
25.	Porąbka-Klimontów	43.3	2.5	–	25	10
26.	Powstańców Śląskich	107.3	5.7	3	51	33
27.	Pstrowski[c]	206.7	3.0	2	27	2
28.	Rozbark	104.6	8.0	18	72	46
29.	Saturn	24.5	1.2	–	11	4
30.	Siemianowice	143.2	4.3	1	39	15
31.	Siersza	67.6	1.6	–	14	13
32.	Sosnowiec	34.4	1.6	1	14	3

(continued)

Table 7.3 (continued)

No.	Mine	Subsidence volume to 2009 (mln m^3)[a]	Average surface lowering in the years 1883–1993 (m)	Anthropogenic denudation rate (mm/year)[b]		
				To 1882	1883–1993	1994–2009
33.	Staszic	61.3	2.2	–	64	103
34.	Szombierki	111.5	10.7	5	97	–
35.	Śląsk-Matylda	60.1	8.0	23	88	–
36.	Śląsk	32.8	1.3	–	66	59
37.	Wawel	135.1	10.6	17	95	35
38.	Wesoła	101.4	1.5	–	19	41
39.	Wieczorek	99.1	5.2	6	47	37
40.	Wujek	70.4	5.1	–	54	72
41.	Ziemowit	138.6	1.6	–	16	42
	Silesian Upland	3379.6	4.4	1–3	28	26
Racibórz-Oświęcim Basin						
42.	1 Maja	34.8	0.9	–	23	23
43.	Anna	81.9	2.3	–	79	83
44.	Borynia	47.3	1.5	–	67	75
45.	Brzeszcze	85.1	2.4	–	28	48
46.	Budryk	23.7	–	–	–	60
47.	Chwałowice	72.6	2.4	–	27	71
48.	Czeczott	22.6	0.6	–	57	82
49.	Dębieńsko[c]	58.9	1.1	–	12	18
50.	Gliwice	34.5	0.3	–	4	6
51.	Janina[c]	68.7	0.8	–	10	20
52.	Jankowice	95.7	4.1	–	53	131
53.	Jastrzębie	59.0	2.0	–	65	89
54.	Knurów	106.1	2.1	–	24	42
55.	Krupiński	25.1	0.2	–	25	44
56.	Makoszowy	119.1	3.3	–	38	55
57.	Marcel	86.5	1.1	–	10	27
58.	Morcinek	6.1	0.1	–	16	31
59.	Moszczenica	35.6	2.2	–	79	47
60.	Piast[c]	91.6	1.0	–	56	55
61.	Pniówek	59.0	0.9	–	47	67
62.	Rydułtowy	99.5	1.7	1	15	25
63.	Rymer	42.7	2.1	–	22	28
64.	Silesia	31.8	1.1	–	14	23
65.	Sośnica	107.3	2.5	–	33	46

(continued)

Table 7.3 (continued)

No.	Mine	Subsidence volume to 2009 (mln m^3)[a]	Average surface lowering in the years 1883–1993 (m)	Anthropogenic denudation rate (mm/year)[b]		
				To 1882	1883–1993	1994–2009
66.	Szczygłowice	76.8	2.4	–	74	69
67.	Zofiówka	54.0	1.9	–	76	84
68.	Żory	5.9	0.3	–	21	13
Racibórz-Oświęcim Basin		1631.9	1.6	1	12	33
Mining area		5011.5	3.3	0.35	21	30

[a]The volume calculated according to the method described in Sect. 1.2

[b]Research periods were established, following the editions of topographic maps used in the morphometric analysis

[c]The mine is located on the border of the Silesian Upland and the Racibórz-Oświęcim Basin

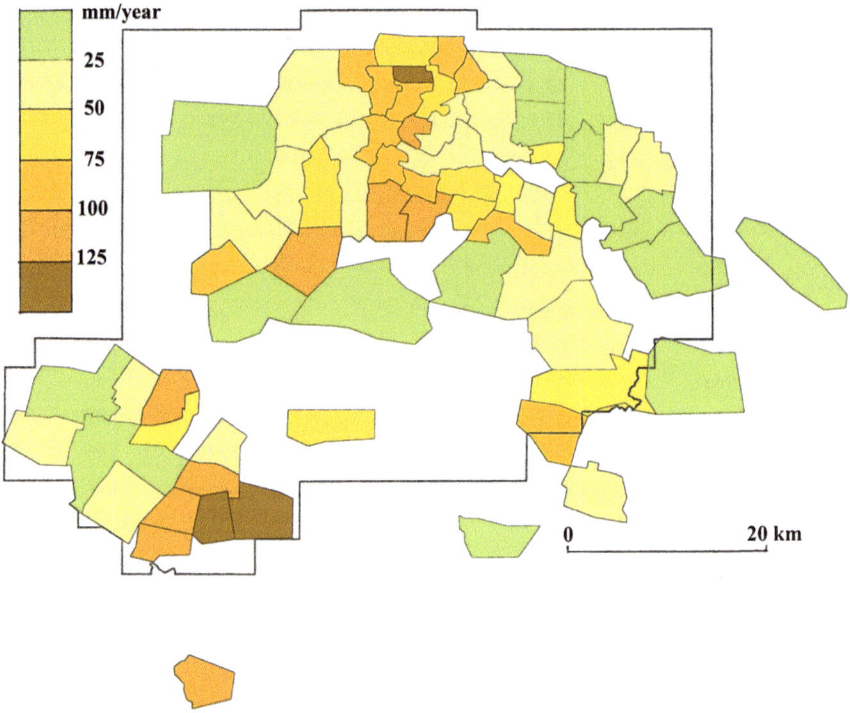

Fig. 7.1 Anthropogenic denudation in the area of mining impacts calculated from coal and waste rocks output to 2009 (mm/year) (from Dulias 2013)

The period until 1882. In the initial period, mining was concentrated in the Silesian Upland. Changes in relief associated with it took place mainly in the area of the Ruda-Chorzów Hills; on the outcrops of Carboniferous rocks, almost 50 % of the coal production took place here, as well as the Siemianowice Upland (about 17 %) and on the Murcki Plateau (nearly 9 %). Transformations of the relief were related to openpit and shallow exploitation in vast areas; hence, the post-mining landscape was poached with small excavations and torn with fissures and sinkholes, between which a number of heaps of waste material were found. The scale of discontinuous deformations was very high because shallow mining resulted in collapse reaching the surface. In the areas of the most intense mining, subsidence and fissures on the relief were so numerous that they forced the storage of fine, pyrophoric carbon in areas preventing the transmission of fire to underground mining excavations, but smoke oozing from underground was a characteristic element of the landscape (Kossuth 1961). Surface waters flowed directly through sinkholes to the mines; in order to reduce the water threat, these forms were dammed or filled up and the fissures sealed (Jaros 1962). Careless casing of the declines also led to collapse; examples of particularly violent collapses of excavations occurred due to the use of the checkerboard exploitation method. The roof supported by pillars was not to break down, which was supposed to protect the surface against mining damage. It turned out, however, that such a method of extraction led to extremely rapid collapse processes; the collapse of rock masses in the mid-nineteenth century was the cause of half of all accidents in the mining industry. One example of such an accident was a sudden and total collapse of a field of 17,000 m^2 within the Król mine (Kossuth 1965). The sizes of some sinkholes were enormous: in the mid-nineteenth century in the Siemianowice mine ("Huta Laura"), a vast heap containing 40,000 tonnes of burning waste collapsed onto underground workings (Piernikarczyk 1933/1934). The Upper Silesian Association of Mining and Metallurgy recommended using the checkerboard method only exceptionally, but this recommendation was not followed because such exploitation achieved fast and high profits, despite the devastation of the relief and an enormous loss of coal deposits.

In the eighteenth and nineteenth centuries, operations were mostly carried out outside built-up areas and mining damage primarily prevented the agriculture of crops because of the "wavy" relief (Jaros 1965). Mining authorities sometimes did not grant licenses for the opening of mines only due to the threat of the destruction of farmland, such as in the area of Mysłowice (Piernikarczyk 1933/1934). Surface deformations became very cumbersome with the growing density of industrial plants and residential development. As time passed, most anthropogenic forms of that period were levelled. Nonetheless, the former areas of shallow mining still posed a threat of unexpected reactivation of old abandoned workings. Examples are the numerous sinkholes that formed in 2010 in Orzesze, above the old mine workings of the Bolesław Śmiały mine (Grygierczyk 2010).

Until 1850, the area of coal mining permits was 232 km^2 in the district of Upper Silesia, less than 12 km^2 in the Cracow district, and about 25–30 km^2 in the Dąbrowa district. The total area of mining permits in the Upper Silesian Basin did not exceed 300 km^2; however, for the most part, these fields were not exploited (Jaros 1965). According to estimates for 1882, mining areas increased to about 400 km^2, including an area of direct mining impact of 200 km^2. On the basis of the calculated theoretical volume of subsidence (68.6 million m^3), it can be deduced that an average lowering of the surface could have been 0.17–0.34 m. The rate of anthropogenic denudation for the period 1769–1882 would be in the range of 1–3 mm/year; the highest values of 23–25 m/year occurred in the Nowy Wirek and Śląsk-Matylda mines, with 16–18 mm/year in the Rozbark, Wawel, and Barbara-Chorzów mines. However, given the fact that almost all sinkholes caused by mining in the eighteenth and nineteenth centuries were levelled and some mining excavations from that period did not reactivate, these values should be considered as hypothetical. It seems that there is no method that obtains reliable results for this mining period.

The period from 1883 to 1993. In the primary period of coal mining, the mining area increased rapidly. In 1913, it already amounted to about 1,100 km^2, including 740 km^2 in the Upper Silesian district, 250 km^2 in the Dąbrowa district, and 110 km^2 in the Cracow district—(Jaros 1965). Mining damages covered more and more area and resulted from various causes. At the turn of the nineteenth and twentieth centuries, and in some areas until the mid-twentieth century, discontinuous deformations formed as a result of the continued shallow operation, as well as by reactivation of old abandoned workings. With the increasing depth of mining works, subsidence troughs appeared in the relief. Despite the increasing use of stowing, continuous deformations covered hundreds of square kilometres, leading to the formation of numerous troughs, with depths locally exceeding up to 30 m (Solarski and Pradela 2010). The formation of mining damages intensified since the 1950s and 1960s, when the exploitation of protective pillars was initiated under cities. Combined with excessive exploitation plans within the framework of the socialist economy, it resulted in surface subsidence on a scale not seen anywhere else in Poland and one of the largest in Europe. In the 1980s, the exploitation of protective pillars still accounted for about one-third of the overall extraction and almost 60 % was conducted using the roof-collapse method, which is especially detrimental to the relief and its development.

In the period of 1883–1993, the volume of anthropogenic denudation significantly increased: on average up to 21 mm/year, with its maximum value of 159 mm/year in the area of the Centrum mine. High values were also obtained for the Szombierki and Wawel mines, at 95–97 mm/year. Much higher values of the D_A rate were obtained by calculating it not for the whole mining area but for the area of mining impact (Fig. 7.1). Some mines conducted operations almost under their whole mining areas, whereas others used larger or smaller parts. For 10 mines, the obtained values reached above 100 mm/year: Centrum, 163 mm/year; Moszczenica and Zofiówka, 120 mm/year; and Śląsk, Śląsk-Matylda, Wawel, Szombierki, Szczygłowice, Jastrzębie, and Pniówek were 102–112 mm/year. The spatial distribution of the D_A rate shows that it was the most important in the Miechowice and Siemianowice Uplands, the Ruda-Chorzów Hills, and the eastern parts of the Tarnowskie Góry and Rybnik Plateaus.

The period from 1994 to 2009. During this period (16 years), many coal mines were liquidated, mainly in the northern part of the USCB; the mining "centre of gravity" shifted to the south and south-west. Today, almost the entire operation is carried out by the roof-collapse method. On unused mining fields, continuous deformation stabilized (expired), and from time to time discontinuous deformations appear. In terms of the D_A, the leader was still the Centrum mine with 154 mm/year, but high values were also achieved in Jankowice (131 mm/year) and Staszic and Pokój (102–103 mm/year). The anthropogenic denudation rate calculated for the area of mining impact reached the highest values for the Jankowice mine (204 mm/year), Centrum (158 mm/year), Pniówek (154 mm/year), and Jastrzębie (149 mm/year). Of the 11 mines for which the anthropogenic denudation rate was greater than 100 mm/year, as many as 9 are located in the Racibórz-Oświęcim Basin, including 5 on the Rybnik Plateau.

Two facts need to be highlighted in the above analysis. Firstly, mining from the period 1769–1882 (114 years), besides posing a potential threat of discontinuous deformations, basically did not leave behind any significant changes in the relief of the USCB, as most forms were levelled or incorporated into the younger forms. Anthropogenic denudation was insignificant. Secondly, coal mining centres, after a long period of strong concentration in the Silesian Upland, are currently "shifting" to the Racibórz-Oświęcim Basin and the anthropogenic denudation rates are increasing in this direction.

The presented calculations of anthropogenic denudation are theoretical. They take into account a fixed percentage of waste rock mining per 1 tonne of coal (20 %), the average rate of hydraulic stowing application for mines affiliated in a given coal union, and the same weight for all types of the extracted waste rock. This means that the results are approximate: for some mines they may be overstated, for others they are too low, and for some of them they are similar to the actual state. Close or too low values (compared with the results obtained with the reliable morphometric method) are obtained primarily for the mines situated within the Miocene zone and in less urbanized areas, while clearly overstated rates were obtained for the old mines located in the area of Carboniferous outcrops, in the most intensely urbanized centre of the Upper Silesian agglomeration, and for some mines located in the Triassic zone. The most divergent values refer to the mines of Katowice, Kleofas, Polska, Szombierki, Wawel, Wujek, Centrum, and Czeladź-Milowice. It seems, therefore, that the above-mentioned imperfections of the method (the use of average calculation indicators for all mines) is complemented with disregard of the volume of soil mass and mining and industrial waste transferred in order to level the surface. This was caused by a lack of relevant statistical data for the entire mining period; only the volume of the contemporary spoil tips is known, but we do not know the volume of ground masses transferred during the expansion of cities. Map analyses of the area of the Katowice conurbation indicate considerable thickness of anthropogenic soil in most built-up parts of cities (Atlas geologiczno-inżynierski 2005). Anthropogenic aggradation can therefore balance or even exceed anthropogenic denudation; its underestimated role in the denudation balance of mining areas was already pointed out by Kupka et al. (2008). At this

Table 7.4 Anthropogenic denudation in the Upper Silesian Coal Basin calculated from the stowing sands extraction (from Dulias 2013)

Sandpit	Area (km^2)	Volume (mln m^3)	Years of mining activity	Average surface lowering (m)	Anthropogenic denudation rate (mm/year)
Szczakowa (5 excavations)	32.39	651.6	56	20.1	359
Dzierżno Duże	6.20	111.0	57	17.9	314
Kuźnica Warężyńska	7.73	82.1	39	10.6	272
Dziećkowice	7.02	64.6	15	9.2	613
Maczki Bór (Western and Eastern)	4.09	52.3	50	12.8	256
Pogoria III	2.37	17.7	13	7.5	574
Dzierżno Małe	1.60	14.0	15	8.8	583
Rogoźnik	1.92	7.6	12	4.0	330
Pogoria I	0.95	6.2	22	6.5	297
Pogoria II	0.83	3.9	21	4.7	224
Betoniarnia	0.27	1.7	34	6.3	185
Jęzor -Wysoki Brzeg	1.76	1.6	19	0.9	48
All sandpits in the USCB	76.27	1032.2	102	13.6	133

point, it is worth repeating the previously quoted remarks that even highly specialized computer programmes, taking into account many natural and mining parameters, do not provide fully reliable forecasts of surface deformation (Popiołek and Ostrowski 1981; Mielimąka 2006; Hejmanowski and Malinowska 2009; Hejmanowski and Kwinta 2010).

The second mineral raw material in terms of the volume of extraction in the USCB is stowing sands. Anthropogenic denudation indicators in the areas of the largest sandpits are very high and amount to several hundred millimetres per year (Table 7.4). This is mostly due to the fact that operation is conducted within poorly resistant material with the open pit method, which is easier and faster compared to the underground method, as already highlighted by Jania (1983).

The highest values of anthropogenic denudation were calculated for the Dziećkowice sandpit (613 mm/year), Dzierżno Małe (583 mm/year), and Pogoria III (574 mm/year), in which operation was carried out on a large and moderate scale, but on a short term (13–15 years). Anthropogenic denudation in the sandpits of the USCB reaches an average value of 133 mm/year, which is 11 times higher than in the area of all coal mines (12 mm/year).

7.2 Anthropogenic Denudation Rates Calculated from Morphometric Analysis

The sizes of anthropogenic denudation presented in this chapter were calculated from the difference between digital elevation models for 1883 and 1993. The method does have its drawbacks, which imposes certain interpretation restrictions (see Sect. 1.2). At the core of its shortcomings is the use of cartographic materials on different scales (1:25,000 and 1:10,000) and therefore different generalization of contour lines. To estimate the altitude error, 10 geomorphological units and 18 catchments located entirely or in a large part out of reach of mining operations were selected, and therefore in theory, characterized by slight changes of relative heights in the period 1883–1993 (Table 7.5). For each of these units, an average height was

Table 7.5 Changes in average heights of selected geomorphological units and catchments located mostly beyond the range of mining activity (from Dulias 2013)

Geomorphological unit/catchment	Area (km^2)	Changes in average height (m)	Geomorphological unit/catchment	Area (km^2)	Changes in average height (m)
Gostynia Plain	207.0	−0.05	Upper Tychy Stream	7.4	−0.01
Cielmice Hummock	3.4	−0.14	Nowe Tychy Stream	5.7	−0.09
Paprocany Hummock	10.4	−0.05	Tributary from Piotrowice	2.0	+0.59
Tychy High Plain	29.9	−0.12	Tributary from Goj	3.9	−0.54
Golejów High Plain	229.0	−0.20	Upper Korzeniec	11.5	+0.14
Pszczyna High Plain[a]	34.3	−0.11	Tributary from Wilcze Gardło	4.8	+0.26
Ząbkowice Hummock[a]	35.8	−0.01	Tributary from Kokoszyce	5.3	−0.09
Gołonóg Hills	11.5	+0.05	Grzybowice Stream	10.1	−0.14
Łagisza Stream Depression	6.0	+0.09	Tributary from Górniki	6.7	−0.19
Tributary from Mikołów	8.4	+0.22	Jamki	14.6	−0.18
Jamna	23.2	−0.26	Tributary from Chudów[b]	3.2	+0.10
Kaskadnik	2.2	−0.23	Stream from Solarnia[b]	6.2	+0.07
Bielawka	3.2	−0.14	Bujaków Stream[b]	0.3	−0.16

[a]Area within the boundaries of the study area

[b]In 1993, the catchment area was outside of mining activity, because the Budryk mine, within which it is now, was established in 1994

calculated for 1883 and 1993. It turned out that in the period of 111 years, these values changed in the range from 0.54 to 0.5 m, with an average of ±0.16 m. On this basis, the error in the values of the denudation indicator for the period 1883–1993 was calculated at 1–2 mm/year.

Anthropogenic denudation was calculated according to the method described in Chap. 1.2. The received values were very diverse, ranging between 2 and 43 mm/year. The highest values of D_A are for the Ruda Hills (43 mm/year), the Siemianowice Upland (40 mm/year), and the Miechowice Upland (33 mm/year) and are undoubtedly related to the morphological effects of the intensive and long-term extraction of coal. The values of 22–26 mm/year were obtained for 5 geomorphological units: the Murcki Plateau, the Chorzów and Kochłowice Hills, the Mysłowice Basin and the Chrzanów Graben. In the latter two units, a significant impact on the value of the D_A indicator was made by the extraction of stowing sands, in addition to coal. The geomorphological effects of sand mining were also reflected in the values of anthropogenic denudation of the Biskupi Bór Basin (15 mm/year), while the D_A rate for the Rogoźnik Hills was 19 mm/year, which reflects the rate of surface lowering both due to coal mining and limestone exploitation. The lowest values of D_A refer to geomorphological units only partially contained within the mining operation: the Wilcza High Plain (2 mm/year), the Czechowice High Plain (4 mm/year), and the Kłodnica Graben (5 mm/year). A relatively small value of the D_A rate (12 mm/year) was obtained for the Rybnik Plateau, which is characterized by intense mining activity. This is due to the fact that for all analysed geomorphological units, an identical time of mining operations was applied, which was 111 years (1983–1993); half the mines in the Rybnik Plateau were established in the postwar period and until 1993 had a maximum of 40 years of coal exploitation. Taking into account this fact, it was calculated that the eastern part of the Rybnik Plateau, in which the mentioned mines are located, has an average D_A value of 57 mm/year. A relatively low rate of anthropogenic denudation (8 mm/year) was also calculated for the Rachowice High Plain, although its eastern portion is within a very intensive mining subsidence. In this case, it is due to the conversion to a very large area (more than 240 km^2). Approximately, the D_A rate is twice the size in this area, amounting to several millimeters per year.

Indicators of anthropogenic aggradation for the researched geomorphological units are significantly smaller compared with denudation rates; they are in the range of 1–9 mm/year, averaging 4 mm/year (Table 7.6). In some areas, the values of A_A are within the margin of error (±1 to 2 mm/year) for the Kochłowice Hills, the Chrzanów Graben, and the Rachowice High Plain; for most units, if taking into account the error, A_A indicators would amount to 1–2 mm/year. The highest values of anthropogenic aggradation, reaching 6–9 mm/year, were obtained for geomorphological units with large quantities of mine tailings accumulated on spoil tips—namely for the Rybnik Plateau, the Mysłowice Basin, and the Kłodnica Graben. The positive denudation balance of +1 to 3 mm/year was calculated only for 4 out of the 25 analysed geomorphological units.

Table 7.6 Anthropogenic denudation balance of selected geomorphological units in the Upper Silesian Coal Basin for the period between 1883 and 1993, calculated from morphometric analysis (from Dulias 2013)

Geomorphological unit	Area[a] (km^2)	Anthropogenic denudation (mm/year)	Anthropogenic aggradation (mm/year)	Anthropogenic denudation balance (mm/year)
Carboniferous zone				
Ruda Hills	54.1	43	3	−40
Chorzów Hills	45.9	26	4	−22
Kochłowice Hills	49.0	24	1	−23
Murcki Plateau	99.8	24	3	−21
Western Mikołów Hummock	30.9	7	4	−3
Eastern Mikołów Hummock	95.5	8	4	−4
Mysłowice Basin	85.0	26	6	−20
Biskupi Bór Basin[2]	54.7	15	3	−12
Triassic zone				
Miechowice Upland	107.8	33	3	−30
Siemianowice Upland	79.1	40	3	−37
Czeladź Upland	25.4	15	3	−12
Bobrowniki Hills	6.6	6	9	+3
Rogoźnik Hills	8.4	19	4	−15
Wojkowice Hummock	3.6	16	8	−8
Miocene zone				
Kłodnica Graben	54.0	5	8	+3
Mleczna Basin	67.2	6	9	+3
Southern Podstokowa Zone	51.8	9	5	−4
Czechowice High Plain	74.3	4	5	+1
Rachowice High Plain	243.6	8	2	−6
Wilcza High Plain	68.7	2	–	−2
Rybnik Plateau	439.9	12	9	−3
Chrzanów Graben	73.2	22	1	−21

(continued)

Table 7.6 (continued)

Geomorphological unit	Area[a] (km^2)	Anthropogenic denudation (mm/year)	Anthropogenic aggradation (mm/year)	Anthropogenic denudation balance (mm/year)
Different geological zones				
Grodziec Elevations	10.6	14	5	−9
Northern Podstokowa Zone	67.0	12	4	−8
Dańdówka Plateau	42.8	12	4	−8

[a]The study area covers only a part of the following geomorphological units: the Wilcza High Plain (70 %), the Rachowice High Plain (94 %), the Chrzanów Graben (77 %), the Rybnik Plateau (77 %) and the Biskupi Bór Basin (41 %)

The calculation of denudation and anthropogenic aggradation in certain areas is subject to errors resulting from the previously highlighted fact that waste rock extracted from underground mines was often stored in concave anthropogenic forms—post-exploitation holes and subsidence troughs. If the formation of a concave and its backfill took place after 1883 and before 1993, the above changes in the relief were not captured in the contour lines of the maps used in the analysis. Thus, the corresponding loss/increase of rock masses was not included in the denudation balance.

The D_A and A_A rates presented above are average values for entire geomorphological units, often very large in terms of their area. In fact, the distribution of values of these indicators in the area of individual units is varied, as illustrated by the denudation balance calculations for smaller catchments within them. In the area of the Miechowice Upland, the denudation balance for selected catchments was as follows: the Rokitnica Stream, 5 mm/year; the Mikulczyce Stream, 25 mm/year; the tributary in Mikulczyce, 51 mm/year, on the Murcki Plateau; the Matownik, 3 mm/year; the Upper Ślepiotka, 11 mm/year; South Bolina, 54 mm/year, on the Rachowice High Plain; the Ostropka, 2 mm/year; and the tributary from Przyszowice, 53 mm/year. The highest negative denudation balance value was obtained for the catchment area of the Orzeł Biały Ditch on the Siemianowice Upland (80 mm/year). For the other catchments in this geomorphological unit—the Michałkowice Ditch, the Dąbrówka Wielka Ditch and the Upper Bytomka, it amounted to 31–34 mm/year.

In the mesoregional scale, the highest rate of anthropogenic denudation was in the Southern Silesian Upland (21 mm/year); the highest rate of anthropogenic aggradation was in the Oświęcim Basin (6 mm/year). The denudation balance of all mesoregions is negative: by far the largest is in the Southern Silesian Upland (17 mm/year) and the lowest in the Oświęcim Basin (2 mm/year; Fig. 7.2).

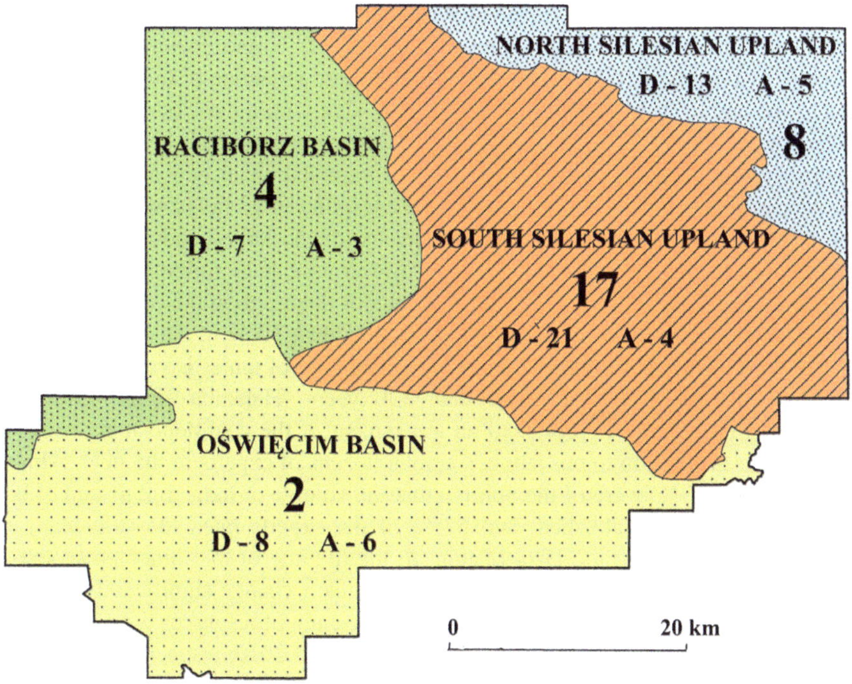

Fig. 7.2 Denudation balance of geomorphological mesoregions in the study area (mm/year) for the period between 1883 and 1993 calculated from morphometric analysis. (**D**) anthropogenic denudation rate, mm/year, (**A**) anthropogenic aggradation, mm/year; *number in large bold font* denudation balance, mm/year

7.3 Forecasts of Anthropogenic Denudation

Mining activities in the Upper Silesian Coal Basin have not been completed. The contemporary mining of mineral resources in the USCB, however, is very limited compared to the intensive production in the late 1970s: 3 times greater for coal mining and 7 times for stowing sands. The exploitation of zinc and lead ores within the USCB has been totally abandoned. In 2010, the largest output of coal mines was reported in the mines of Piast (4.2 million tonnes), Borynia and Ziemowit (3.8 million tonnes each), and Murcki-Staszic and Knurów-Szczygłowice (3.7 million tonnes each). In 2014, the mines with the largest output included Piast (3.3 million tonnes), Ziemowit (3.1 million tonnes), Pniówek (2.8 million tonnes), and Budryk (2.7 million tonnes). Mining has left most of the Triassic geological zone. At present, about 7 % of coal output in the USCB comes from this zone, while from the Carboniferous zone it is about 30 %. The Miocene zone has the largest share of coal mining at present—about 65 %. In the last two decades, the share of the Racibórz-Oświęcim Basin in coal production has already reached 52 %.

The introduction of new criteria for the coal deposits (a minimum thickness of seams of 1 m, a total sulphur content of less than 2 %, and documentation of the deposit to a depth of 1000 m) has led to a significant reduction of the resource base. As of the end of 2014, a total of 136 deposits were within the Upper Silesian Coal Basin, with a total balance of resources estimated at less than 42 billion tonnes. Industrial resources, which are intended for exploitation in deposit development projects, amount to 3.44 billion tonnes and are found in 52 fields, within the limits of the 28 currently active mines. The mines with the largest industrial resources include Janina (354 million tonnes) and Chwałowice (213 million tonnes). Most industrial resources, which is about two-thirds of them, are in the Miocene zone, a quarter in the Carboniferous zone, and only less than 10 % in the Triassic zone.

The sufficiency of resources and the associated life of mines is variously defined and ranges from a few to more than 130 years (Halemba-Wirek, Zofiówka, Budryk). In 2002, Kicki presented an assessment of the coal deposit economy in Poland, based on anonymous surveys among dozens of experts in the field. Almost half of them (45 %) were of the opinion that the volume of industrial resources had not been properly established, and 57 % of the experts rated the sufficiency of resources for 15–35 years. Data relating to the lifetime of the mines presented in various works (mining strategies, access plans, etc.) change significantly, even at intervals of several years. For example, the sufficiency of resources was determined for the following mines for the years 2003, 2007, and 2010, respectively: Knurów, 44, 38, and 52 years; Piast, 29, 37, and 57 years; and Budryk, 89, 72, and 130 years, respectively. In this work, mines were divided into two age groups—those with prospects for activity for up to 30 years and over 30 years—thus combining the forecast of mine life according to Paszcza and Borsucki (2010) with the ones presented in 2007 in the *Strategy for Coal Mining for 2007–2015* (Strategia dla górnictwa 2007) (Fig. 7.3). According to the estimates given in the first of these studies, the mines will be open 1.5–2 times longer than assumed in the quoted strategy.

Analysis of the possibilities for redevelopment of liquidated coal mines in the USCB was provided by Jureczka and Galos (2007), concluding that they are real in the case of deposits within 8 mines: Morcinek, Siersza, Żory, Niwka-Modrzejów, Moszczenica, Dębieńsko, Jan-Kanty, and Czeczott. Most of these deposits (5) are located in the Miocene zone. The possibility of mine reactivation has already been applied: the resumption of coal extractions from the most prospective deposits in the area of the liquidated Dębieńsko mine is planned for 2017. Work on facilitating a new deposit is also under way in Bzie Dębina West for the Zofiówka mine; operations are scheduled to begin in 2018–2019.

Contemporary geomorphological processes in the study area occur in the natural environmental conditions clearly changed due to long-term and intense economic activity. Changes in morphometric parameters of the relief and lithology of substrate, water regimes, and land use in open and closed systems were characterized in Chaps. 5 and 6. On the basis of the above-presented prospects for the development of coal mining, it may be stated that for at least a few decades it will continue to have an impact on the lithosphere and hydrosphere of the USCB (approximately

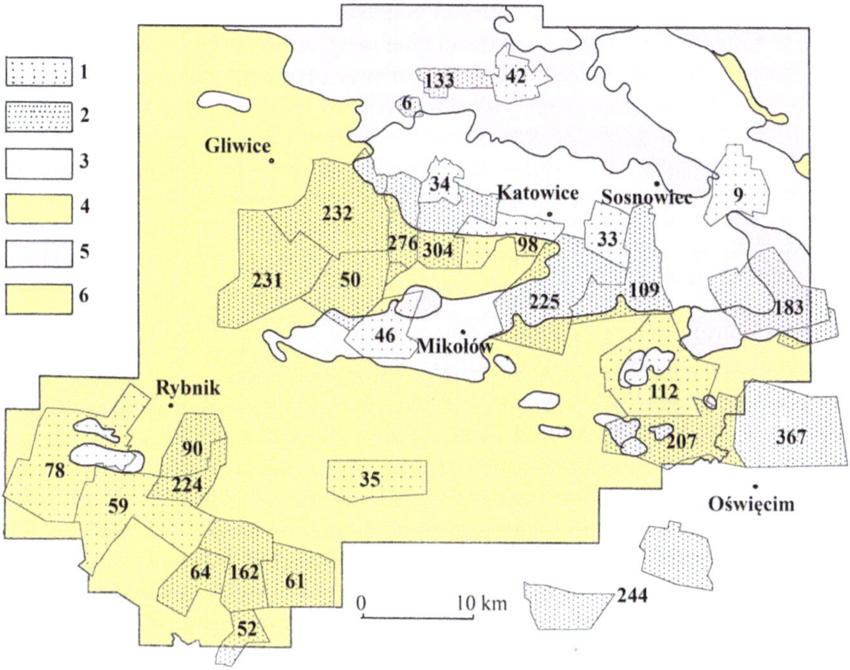

Fig. 7.3 Coal resources and forecasted life of coal mines in the Upper Silesia Coal Basin against a background of geological zones (based on Paszcza and Borsucki 2010, Strategia dla górnictwa 2007). Life of coal mines: (**1**) to 30 years, (**2**) more than 30 years, (**3**) Permian rocks, (**4**) Miocene zone, (**5**) Triassic zone, (**6**) Carboniferous zone. The *number in bold* indicates the resources in Mt

1,000 km^2). The largest development prospects for coal mining are expected in the deposits in the area of the Rachowice High Plain, the Podstokowa Zone, the Rybnik Plateau, the Vistula Valley, the Murcki Plateau, and western parts of Chorzów and Kochłowice Hills. Thus, mining is descending from the Silesian Upland to the surrounding Racibórz and Oświęcim Basins and will affect the areas of generally lower relative heights, with the exception of the Rybnik Plateau. Bearing in mind the types of relief modelling distinguished by Bogacki and Starkel (1999), it may be said that the morphodynamic character of geomorphological units, within which underground extraction of coal will be conducted despite the relatively low relative heights, is predominantly degradative. A greater part of the mining areas are, in fact, situated on plateaus, uplands, hummocks, and tectonic hills.

The height and distribution of precipitation over the course of a year—and in particular, slopewashing—are going to have an impact on the course of contemporary geomorphological processes. In the USCB, annual rainfall ranges from 628 mm in Chałupki in the Oder valley to up to 842 mm in Katowice-Murcki, which makes it quite varied (214 mm). Based on data from dozens of precipitation stations located within the study area and in the immediate vicinity, it may be said that the area with the least rainfall is the Racibórz Basin, with usually about

670 mm per year. The second area of relatively low rainfall extends from the Czeladź Upland to the Mysłowice Basin. The most rainfall is observed in the highly elevated Mikołów Hummock, the southern part of the Murcki Plateau, and the latitudinal belt stretching along the central part of the Oświęcim Valley, with an average rainfall of 810 mm. Significant differences in terms of rainfall erosivity between potential mining regions result from the above, especially between the Rachowice High Plain and the Murcki Plateau.

The area of the Triassic geological zone, which has the smallest amount of industrial resources of coal, is subject to the weakest mining anthropopressure in the whole USCB. In 2010, mining operations were conducted here in the Bobrek-Centrum mine (2.1 million tonnes in 2010), ZG Piekary (1.5 million tonnes), the Siltech company (0.2 million tonnes), and partly by ZGE Sobieski Jaworzno III. In the long term (over 30 years), operations were to be carried out in the area of the Bobrek-Centrum mine, which is the region of one of the largest, already existing surface lowering; thus, it was predicted that the impact of mining in the eastern part of the Miechowice Upland would be enhanced (Dulias 2013). However, in 2014, as a result of mining restructuring, the operation was only led by Bobrek-Centrum (0.4 million tonnes) and Jaworzno (1.3 million tonnes) mines. The circulation of matter in the Triassic zone will be executed substantially within landlocked basins, as these forms represent more than 26 % of the former mining area. We assume that the landlocked basins, with depths from 1 to 15 m (calculated from the bottom of the form to the lowest point of the watershed), will be filled with sediments to a level restoring river runoff at a rate from 0.2 to 1.7 mm/year (Dulias 2013). Theoretically, this process may last even tens of thousands of years. Re-inclusion of these areas in the fluvial drainage system seems to be very remote in time.

In the Carboniferous zone, the main industrial resources are located on the Murcki Plateau (almost 367 million tonnes), the Kochłowice Hills, and the Mysłowice Basin. The lifetime of mines in some forecasts is determined to be more than 60 years. Except for the already-mentioned Halemba-Wirek mine, there are mines of the Mysłowice-Wesoła (80 years) and the Murcki-Staszic (60 years) (Paszcza and Borsucki 2010). Assuming that an average rate of anthropogenic denudation caused by underground mining (estimated on the basis of the balance of output in the years 1994–2009) in the Silesian Upland was 25 mm/year, the surface lowering after 60 years of operation may be approximately 1.5 m. The maximum surface lowering on the Murcki Plateau by the end of the mining license will reach 5–9 m in some places (Atlas geologiczno-inżynierski 2005). The landlocked basins that currently exist in the Carboniferous zone represent 4.5 % of the mining area, but almost all forms are beyond the range of contemporary mining operations and will gradually be filled with material coming from the denudation of slopes. The rate of denudation, compared with the Triassic zone, is likely to be lower as many basins are extensively built-up and included in municipal sewage systems. The circulation of matter in the remaining part of the Carboniferous zone will take place in the context of open fluvial systems with significantly changed circumstances—

changes in declines, lowered base level of erosion, increased inclination of slopes, and increased flows resulting from "alien" water supply.

The Miocene geological zone is found within the most intense mining activity. More than 60 % of the industrial resources of coal in the USCB are present here in the bedrock of three different and remote geomorphological areas: (1) the Rachowice High Plain and the Podstokowa Zone, (2) the Rybnik Plateau, and (3) the Valley of the Upper Vistula and its surrounding tectonic units of the Lędziny and Libąż Hills, the Bieruń Hummock and the Chrzanów Graben. It is predicted that most of the mines will be open a few decades, with the exception of old plants in the western part of the Rybnik Plateau and the mines of Krupiński, Bolesław Śmiały, and Ziemowit. Assuming about 50 years of mining activity and the rate of denudation estimated for the area of the Racibórz-Oświęcim Basin at 37 mm/year (based on the balance of output in the years 1994 to 2009), it may be predicted that the surface will lower by an average of almost 2 m. The rate of landlocked basins (covering less than 3 % of the Miocene zone) filling with sediments will vary—it will be the highest on the Rybnik Plateau because many forms were created by damming deep valleys of highly inclined slopes with spoil tips. A significant share of anthropogenic aggradation will therefore be the deposition of material from spoil tips. The processes of fluvial erosion and accumulation are going to be rather intense, especially in the Rachowice High Plain and the Podstokowa Zone. If the rate of fluvial erosion in some sections of riverbeds continues to be comparable to the present ratio, which is in the range of 0.5–24 cm/year, it may be predicted that the beds will deepen from 1.5 to as much as 7 m (Dulias 2013). Mining in the Miocene zone involves substantial extraction of waste rock; therefore, further built-up of old spoil tips and the formation of new ones is to be expected. Movement of rock masses, frequently on a large scale, is also related to the reclamation of mining areas (Skawina 1968; Skawina and Trafas 1971; Greszta and Morawski 1972). These targeted actions—which were planned even before the onset of mining activity and conducted during its operation, as well as after the completion of exploitation—significantly shape the rate of denudation/anthropogenic aggradation. In many places of the USCB, the process of "dismantling" many old spoil tips is taking place and the waste heap material is used for road construction.

The underground extraction of coal has modified the original dynamic balance of the rock mass. In the areas of liquidated coal mines, this balance currently remains unstable until the moment of activation of an impulse, which is not necessarily connected with mining (e.g. heavy rain), causing a complete or an incomplete discharge of the pressure in the rock mass (Niedojadło 2005). Discontinuous deformations—but also the "residual" continuous deformations—may be revealed many years after the cessation of mining activities. The potential anthropogenic denudation associated with the reactivation of old workings does not seem to have a major impact on the total volume of denudation.

The cessation of mining operations and the abandonment of drainage leading to flooding of mines are associated with the rising water table and the restoration of a hydrodynamic balance in the rock mass (Wilk 2003; Frolik 2005; Bukowski 2006). The filling of a depression cone lasts about 10 years and may lead to two types of effects: (1) an activation of old workings and emergence of discontinuous deformations and an increase of surface lowering; or (2) an increase in the volume of the rock mass layers with swelling properties and the formation of surface convexities. Niedojadło (2005) gives an example from the Zwickau–Oelsnitz coal mining region: here, 20 years after the cessation of mining, mining damages continue to arise, including surface convexities of 9–15 cm. Most of the mines of the USCB are hydraulically connected to one another, so the concept of a complete cessation of drainage in the case of abandoned coal mines and the restitution of the water table to a state of hydrodynamic balance at the level of a few meters above the surface in areas of greatest subsidence is invalid (Bukowski and Augustyniak 2005). The Siersza and Morcinek mines are isolated and are currently being flooded.

References

Aleshina IN, Snytko VA, Szczypek S (2008) Mining induced ground subsidences as the relief-forming factor on the territory of the Silesian Upland (Southern Poland). Geogr Nat Resour 29:288–291

Aparta M, Jania J (1980) Niektóre zagadnienia antropogenizacji rzeźby na Wyżynie Śląskiej. In: Przeobrażenie środowiska geograficznego w obszarach uprzemysłowionych i zurbanizowanych, Sosnowiec-Kozubnik, pp 27–39

Atlas geologiczno-inżynierski aglomeracji katowickiej 1:10000. Katowice–Warszawa–Wrocław (2005)

Bennett EM, Carpenter SR, Caraco NF (2015) Human impact on erodible phosphorus and eutrophication: a global perspective. Bioscience 51(3):227–234

Bogacki M, Starkel L (1999) Typologia i regionalizacja współczesnych procesów rzeźbotwórczych. In: Starkel L (ed) Geografia Polski. Środowisko przyrodnicze. Wyd Nauk PWN, Warszawa

Bukowski P (2006) Zawodnienie powierzchni terenu spowodowane działalnością górniczą prowadzoną w GZW w okresie rozwoju górnictwa lat. 70. i 80. XX wieku do okresu restrukturyzacji kopalń. Przegl Górn 62, 5:15–24

Bukowski P, Augustyniak I (2005) Analiza zjawisk związanych z zaprzestaniem odwadniania wyrobisk górniczych na przykładzie byłej kopalni Maria. Bezp Pracy Ochr Środ Górn 1(125):13–20

Demek J (1973) Úvod do studia reliefu Země. SPN, Praha

Dębski K (1959) Próba oszacowania denudacji na obszarze Polski. Prace Stud Kom Gosp Wod 2, 1, Warszawa

Dulias R (2005) Krzywe hipsograficzne obszaru osiadań górniczych (na przykładzie okolic Piekar Śląskich). In: Krzemień K, Święchowicz J (eds) Kotarba A. Współczesna ewolucja rzeźby Polski, Kraków, pp 115–120

Dulias R (2007a) Wpływ górnictwa węgla kamiennego na zmiany rzeźby obszaru KWK Miechowice na Wyżynie Śląskiej. Acta Geogr Silesiana 1:5–12

Dulias R (2007b) Geomorfologiczne skutki eksploatacji węgla kamiennego w Zagłębiu Dąbrowskim. Kształ środ geogr ochr przyr obsz uprzem zurb 38:11–22

Dulias R (2010) Anthropogenic denudation in mining areas: a case study of "Andaluzja" mine, Silesian Upland. Anthropogenic Aspects Landsc Transform 6:23–28

Dulias R (2011) Impact of mining subsidence on the relief of the Rybnik Plateau, Poland. Z Geomorph 55, Suppl 1:25–36

Dulias R (2013) Denudacja antropogeniczna na obszarach górniczych na przykładzie Górnośląskiego Zagłębia Węglowego. Wyd Uniw Śląskiego, Katowice

Dulias R, Szczypek T (2005) Badania antropogenicznych basenów sedymentacyjnych Wyżyny Śląskiej z zastosowaniem ^{137}Cs. Geomorfologický sborník 4:15–18

Frolik A (2005) Odwadnianie kopalń likwidowanych. Wiad Górn 7–8:359–368

Greszta J, Morawski S (1972) Rekultywacja nieużytków poprzemysłowych. PWRiL, Warszawa

Grygierczyk K (2010) Zakaz wstępu do lasu! Pojawiły się niebezpieczne zapadliska. Polska Dziennik Zachodni 13(08):2010

Hejmanowski R, Malinowska A (2009) Evaluation of reliability of subsidence prediction based on spatial statistical analysis. Intern J Rock Mech Min Sci 46:432–438

Hejmanowski R, Kwinta A (2010) Modelowanie deformacji ciągłych powierzchni terenu w warunkach zmiennego zalegania złoża. Gosp Sur Min 26(3):143–153

Jania J (1983) Antropogeniczne zmiany rzeźby terenu wschodniej części Wyżyny Śląskiej. Dokumentacja teledetekcyjna. Prace Nauk Uniw Śląskiego 575:69–91

Jaros J (1962) Historia kopalni „Król" w Chorzowie (1791–1945). Katowice

Jaros J (1965) Historia górnictwa węglowego w Zagłębiu Górnośląskim do 1914 roku. Inst. Historii Kult Mater PAN. Zakł Narod Ossolińskich, Wyd PAN, Wrocław-Warszawa-Kraków

Jureczka J, Galos K (2007) Niektóre aspekty ponownego zagospodarowania wybranych złóż zlikwidowanych kopalń węgla kamiennego w Górnośląskim Zagłębiu Węglowym. Polit Energ 10(2):645–662

Kicki J (2002) Gospodarka zasobami złóż węgla kamiennego w Polsce – ich wystarczalność i jej uwarunkowania w oczach ekspertów. Gos Sur Min 28(2):19–36

Kossuth S (1961) Zarys rozwoju techniki górniczej w kopalniach węgla w Zagłębiu Górnośląskim do połowy XIX wieku. Państ Rada Górn 30, E (2) - Górnictwo polskie w tysiącletnim okresie istnienia Państwa Polskiego. Wyd Geol, Warszawa

Kossuth S (1965) Górnictwo węglowe na Górnym Śląsku w połowie XIX wieku. Prace GIG, Wyd Ślask, Katowice

Kozarski S, Rotnicki K (1978) Problemy późnowürmskiego i holoceńskiego rozwoju den dolinnych na Niżu Polskim. PTPN Prace Kom Geogr Geol 2, 4, Poznań

Kupka R, Frolik H, Dulias R (2008) Zmiany rzeźby na obszarze górniczym zlikwidowanej kopalni „Katowice-Kleofas". Informacja ogólna. Kształ środ geogr ochr przyr obsz uprzem zurb 39:26–31

Kupka R, Szczypek T, Wach J (2005) Morphological effect of 200-years long hard coal exploitation in Katowice. In: Szabó J, Morkūnaitė R (eds) Landscapes—nature and man. University of Debrecen, Lithuanian Institute. Geology Geography, Debrecen-Vilnius: 95–100

Łajczak A (2009) Rozmiary transportu zawiesiny w dorzeczu Wisły powyżej Krakowa i zmiany spowodowane działalnoscią człowieka w drugiej połowie XX wieku. Kształ środ geogr ochr przyr obsz uprzem zurb 40:102–121

Madowicz A (2001) Osiadania terenu na obszarze Jastrzębia Zdroju w latach 1974–1997. Kształ środ geogr ochr przyr obsz uprzem zurb 31:15–21

Mielimąka R (2006) Pomierzone i prognozowane krzywizny terenu górniczego na przykładzie obserwacji geodezyjnych z KWK „Budryk". Górn Geol 1(4):81–92

Mirek K, Isakow Z (2009) Preliminary analysis of Inwar data from south-west part of Upper Silesian Coal Basin. Gos Sur Min 25(3):239–246

Niedojadło Z (2005) Ocena i prognozowanie końcowych wpływów eksploatacji w rejonach zlikwidowanych zakładów górniczych. In: Kwiatek J (ed) Problemy eksploatacji górniczej pod terenami zagospodarowanymi, Ustroń: 389–403

Paszcza H, Borsucki D (2010) Prognoza inwestycji w górnictwie węgla kamiennego – horyzont 2015 (2020), Warszawa. www.proinwestycje.pl

Perski Z (2000) Zastosowanie satelitarnej interferometrii radarowej do określania dynamiki i zasięgu górniczych deformacji terenu na przykładzie wybranych obszarów Górnośląskiego Zagłębia Węglowego. Prace WNoZ UŚ 8:9–39

Piernikarczyk J (1933/1934) Historia górnictwa i hutnictwa na Górnym Śląsku. Śląski Zw Akad, Katowice

Popiołek E, Ostrowski J (1981) Próba ustalenia głównych przyczyn rozbieżności prognozowanych i obserwowanych poeksploatacyjnych wskaźników deformacji. Ochr Teren Górn 58

Price SJ, Ford JR, Cooper AH, Neal C (2011) Humans as major geological and geomorphological agents in the Anthropocene: the significance of artificial ground in Great Britain. Phil Trans R Soc A 369:1056–1084

Sinkiewicz M (1998) Rozwój denudacji antropogenicznej w środkowej części Polski Północnej. UMK Toruń

Skawina T (1968) Klasyfikacja terenów pogórniczych dla potrzeb rekultywacyjnych. Ochr Teren Górn 6

Skawina T, Trafas M (1971) Zakres wykorzystania i sposób interpretacji wyników badań geologicznych dla potrzeb rekultywacji. Ochr Teren Górn 16

Słupik J (1973) Zróżnicowanie spływu powierzchniowego na fliszowych stokach górskich. Dokum Geogr 2

Solarski M, Pradela A (2010) Przemiany wybrany form rzeźby Wyżyny Miechowickiej w latach 1883-1994. Z badań wpł antrop środ 11:78–92

Strategia dla górnictwa węgla kamiennego na lata 2007–2015 (2007)

Wilk Z (ed) (2003) Hydrogeologia polskich złóż kopalin i problemy wodne górnictwa. Uczel Wyd Nauk Dydakt, Kraków

Wojciechowski T (2007) Osiadanie powierzchni terenu pod wpływem eksploatacji węgla kamiennego na przykładzie rejonu miasta Knurowa. Przegl Geol 55(7):589–594

Zapletal L (1968) Geneticko-morfologická klasifikace antropogennich forem reliéfu. Acta Univ Palackinae Olomunicesis Fac Rerum Natur 23 Geogr Geol VIII:239–427

Zapletal L (1969) Úvod do antropogenní geomorfologie. Univ Palackého, Olomouc

Żmuda S (1973) Antropogeniczne przeobrażenia środowiska przyrodniczego konurbacji górnośląskiej. PWN, Warszawa–Kraków

Chapter 8
Anthropogenic Denudation Rate in Other Mining Areas

Due to their economic activity, humans have intentionally or unintentionally become a landscape shaping factor. Apart from agriculture, one of the main forms of anthropopressure is open-pit and underground extraction of raw materials. Due to the large scale of lithosphere transformation, humankind is believed to be a geological or geomorphological factor (Nir 1983; Hooke 1994, 2000; Panizza 1996; Wilkinson 2005; Goudie 1993, 2006). The impact of mining activities on the bedrock and relief may be characterized by the volume of transferred material or the denudation or aggradation rates, mostly expressed in millimetres per year. For example, Douglas and Lawson (2001) reported that 0.7–1 billion tonnes of ground material is transported yearly in the United Kingdom as a result of open-pit mining, waste disposal, and the development of cities. Rivas et al. (2006), based on the research of several mining areas in Argentina and Spain, reported the volume of anthropogenic denudation as 0.8 and 2.4 mm/year, respectively. The intensive exploitation of sediments from the Loire riverbed in the Centre-Val de Loire in France, amounting to 6.4 million tonnes per year compared to a six-fold lower natural sediment supply, contributed to the deepening of the bed by 3 m (Gąsowski 1994).

The volume of denudation in mining catchment areas is often estimated using geochemical markers in sediments of rivers and reservoirs, usually based on an increased content of a given metal compared to the geochemical background (Klimek 1996; Müller et al. 1999). The literature also raises the problem of an increased denudation rate in karst areas due to mining activity (e.g. Tyc 1997; Wanfang 1997). Pulido-Bosch et al. (2004) revealed that the intensification of denudation in karst areas in the semi-arid Tabernas-Sorbas basin in Spain has its cause in the exploitation of gypsum, initiated in the 1980s. On the other hand, Sprynskyy et al. (2009) calculated that the rate of anthropogenic chemical denudation, due to the drainage of the sulphur mining area that has been ongoing for the past 29 years in western Ukraine, amounted to 0.062 mm/year (17 952 m^3/km^2), which was almost 80 times higher than the rate of natural denudation. These and dozens of other examples prove that open-pit and underground exploitation of raw materials is reflected in the denudation balance of the areas in which it is conducted.

R. Dulias, *The Impact of Mining on the Landscape*,
Environmental Science and Engineering, DOI 10.1007/978-3-319-29541-1_8

In a geotechnical environment, such as a mining area, the period of intense movement of rock masses generally ends with the cessation of operations; it is resumed within the established anthropogenic forms, but as a result of natural geomorphological and geomechanical processes. Mining anthropopressure is therefore periodic (episodic): it lasts from the onset of works that make the deposit available, through the years of production, until the mine closure and termination of a majority of the mining impact. It is evocatively presented by Skinner (1978) in the analysis of the stages of development of metal ores mining. Initially, with the growing number of discovered deposits, the number of mines increases; then, the smaller mines become exhausted and the pace of exploitation in the remaining main mines exceeds the discovery of new resources. Before a raw material is exploited, its import is initiated, which increases exponentially and expresses the inability of a country to meet the needs in this area. It seems that this model does not apply to all mineral resources because mineral exploitation may be limited, not only due to the exhaustion of resources but also for economic, social, political, or technological reasons. An example is the Polish mining of sulphur, which in just 50 years since the discovery of large deposits experienced both development on a global scale and a drastic decrease in mining due to the spread of sulphur recovery from natural gas and crude oil. The history of zinc and lead ore mining, in turn, fits well with Skinner's model (1978). Due to the depletion of available resources of these ores, Poland has shifted from the group of major zinc producers in the world to the group of importers.

The geomorphological effects of mining and the volume of anthropogenic denudation in the Upper Silesian Coal Basin presented in the previous chapters require a comparison with other mining areas. This chapter provides the characteristics of the mining rock mass movement in Poland as a result of exploiting 12 major mineral resources, as well as an estimate of the intensity of anthropogenic denudation resulting from the mining activity. It also presents the impact of coal mining on the relief of two European basins—the Ruhr Basin and the Ostrava-Karvina Basin.

8.1 Mining Areas in Poland

The history of mining in Poland dates back to the Neolithic Age. In the underground workings in Krzemionki Opatowskie in the Świętokrzyskie Mountains region, striped flint was mined at the time. Traditions of iron, silver, lead, rock salt, and rock material mining date back to at least the beginnings of the Polish state. However, the intensive development of the mining industry began only in the nineteenth century, with the exploitation of coal, zinc and lead ores, and rock materials. After World War II, rich deposits of sulphur and copper were discovered and new basins of coal, zinc and lead ores, stowing sands, and other rock minerals were established (Kozłowski 1983). The intensive mining activities of the last 200 years have changed the landscape of many regions in Poland. According to

Starkel (1988), geosystem changes in Poland had never been as big and important as during and after the industrial revolution. According to Maruszczak (1988), the human conversion of the "closed" forest landscapes into the "open" agricultural ones was done at a faster rate than the transition from the tundra-steppe landscapes into the forest landscapes due to enormous climatic changes at the turn of the Pleistocene and the Holocene. The transformation of agricultural or forest landscapes into mining landscapes happened at an even faster pace.

Mining plays an important role in the Polish economy. In the last two decades, the annual production of minerals ranged from 350 to nearly 440 million tonnes. With the addition of waste rock mining (30–80 million tonnes per year) and rock material mining by individual owners (no statistics), mining production may amount to, according to conservative estimates, over 500 million tonnes per year. For the 1970s, in relation to official statistics, almost twice the volume of extraction was assumed. For example, according to Leszczycki (1980), in 1970, the movement of rock masses affected approximately 700 million tonnes, which was 320 million tonnes more than reported in the official statistics of production. Kozłowski (1983) estimated the extraction of mineral resources on the "grey market" for 1978 as at least at 100 million tonnes. According to Krupiński (1971), in the 1960s, Poland was one of the top ten mining countries in the world. Currently, the role of Poland in the global mining industry has clearly decreased but is still significant.

In Poland, approximately 60 different raw materials are exploited—a few on a large scale. In 2009, mining operations were as follows: mineral raw materials, 257.9 million tonnes; power raw materials, 127.6 million tonnes; metallic ores, 25.5 million tonnes; and chemical raw materials, 3.4 million tonnes. In total, 414.4 million tonnes of minerals were extracted and 50.3 million tonnes of mining waste were produced. The main minerals were natural aggregates (34 % of total extraction), coal (17 %), lignite (13.8 %), dimensional and crushed stones (13.3 %), limestones and marls for the cement and lime industry (8.5 %), copper ores (5.6 %), stowing sands (2.4 %), clayey raw ceramic building materials (1.3 %), rock salt (0.7 %), dolomites (0.7 %), zinc and lead ores (0.6 %), and sulphur (0.07 %). The output of all other minerals provided only 2.03 % of total extraction, so the following analysis only included the 12 aforementioned minerals (Table 8.1). The main mining areas in Poland are presented in Fig. 8.1.

Coal occurs in Poland in three basins—the Upper Silesian Basin, the Lower Silesian Basin, and the Lublin Basin. The first mention of the coal mines in Lower Silesia is from 1434; however, until the eighteenth century, operation was insignificant. The main development of mining began in the first half of the nineteenth century, when 30–50 mines were operating. The Lower Silesian Basin, despite growing extraction, quickly lost its leading role in favour of Upper Silesia. After World War II, there were 6 active mines, the combination of which formed two: the Wałbrzych mine and the Nowa Ruda mine. Lower Silesian mining, therefore, was focused in two areas: the Wałbrzych Basin and the Nowa Ruda Depression in the foreland of the Sowie Mountains, in an area of approximately 530 km^2. Coking coal was mined here from relatively thin, steep seams located at a great depth. In 2000, operations were abandoned due to deteriorating geological and

Table 8.1 Output of major raw materials in Poland during the years 1960–2009 (from Dulias 2013)

Raw materials	Output (mln tonnes)	Percent of the total output
Power raw materials	9349.9	45.4
Coal	6975.3	33.9
Lignite	2374.6	11.5
Metal ores	1197.8	5.8
Zinc and lead ores	213.5	1.0
Copper ores	984.3	4.8
Chemical raw materials	302.8	1.5
Sulphur[a]	126.6	0.6
Rock salt	176.2	0.9
Mineral raw materials	9755.0	47.3
Dolomites for metallurgical industry	144.3	0.7
Limestones for lime and cement industry	1909.1	9.3
Dimension and crushed stones[b]	1200.0	5.8
Clayey raw materials	507.8	2.5
Gravel aggregates	3826.1	18.6
Stowing sands	2167.7	10.5
Total	20605.5	100

[a]Output for the years 1958–2009
[b]Output in the years 1990–2009 amounted to 508 million tonnes; for the years 1960–1989, the output is estimated

mining conditions. The collected statistics show that in the period 1769–2000 in the Lower Silesian Basin, 492.2 million tonnes of coal were extracted, which was only 4.4 % of national production. In contrast to the Upper Silesian Coal Basin, extensive mining (44 % of the total) took place in the first half of the twentieth century, despite the crisis of the inter-war period when 9 mines were liquidated and flooded.

The mining in Lower Silesian led to significant transformations of the relief, especially in the Wałbrzych region in the range of impact of the Wałbrzych, Thorez, and Victoria mines. In the period of 1865–1996, a total of 39 spoil tips were formed here, covering an area of more than 3 km^2, which accumulated nearly 83 million m^3 of waste rock; 8 settling tanks were located on hilltops of dumps with a total area of 0.74 km^2 and a volume of 9 million m^3 (Wójcik 1995, 2006, 2011). The highest spoil tip reached a height of 105 m. The large amount of mining waste accumulated in the Wałbrzych area resulted from the fact that in the 1970s and 1980s, 1.5 tonnes of waste rock was extracted per 1 tonne of extracted coal. Mining subsidence occurred in an area of about 35 km^2—almost reaching up to 18 m in Wałbrzych–Biały Kamień, where a landlocked basin with a water reservoir was formed. In the remaining area, the surface decreased from 0.1 m on the wings of the Wałbrzych Basin to 10 m in its centre, in the area where protective pillars were exploited in the 1980s. The volume of subsidence troughs is estimated at 245 million m^3 (Wójcik

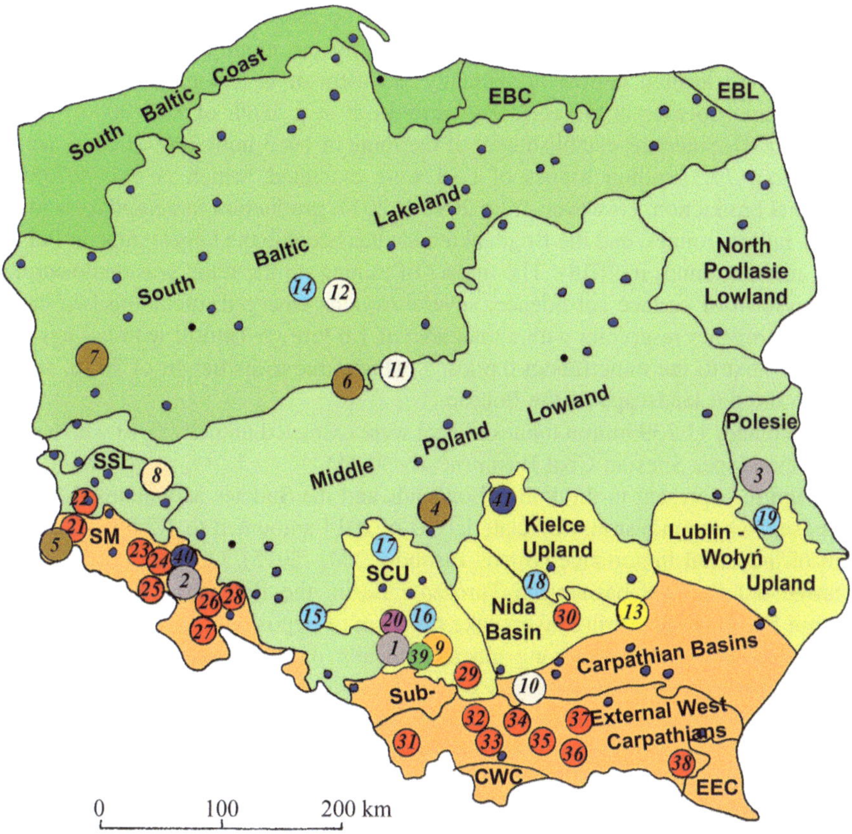

Fig. 8.1 Main mining areas in Poland against a background of geomorphological sub-provinces (from Dulias 2013): **Coal**: (*1*) Upper Silesia, (*2*) Lower Silesia, (*3*)—Lublin, **Lignite**: (*4*) Bełchatów, (*5*) Turoszów, (*6*) Konin, (*7*) Sieniawa, **Copper ores**: (*8*) Legnica-Głogów, **Zinc and lead ores**: (*9*) Silesian-Cracow, **Rock salt**: (*10*) Subcarpathian, (*11*) Kłodawa, (*12*) Inowrocław, **Sulphur**: (*13*) Tarnobrzeg, **Limestones and marls**: (*14*) Kuiavian, (*15*) Opole, (*16*) Zawiercie, (*17*) Częstochowa, (*18*) Kielce, (*19*) Chełm-Rejowiec, **Dolomites**: (*20*) Bytom-Siewierz, **Dimension and crushed stones**: (*21*) Lubań, (*22*) Bolesławiec, (*23*) Złotoryja, (*24*) Strzegom, (*25*) Wałbrzych, (*26*) Pilawa, (*27*) Kłodzko, (*28*) Strzelin, (*29*) Krzeszów, (*30*) Pińczów, (*31*) Bielsko, (*32*) Sułkowice, (*33*) Krzczonów, (*34*) Myślenice, (*35*) Limanowa, (*36*) Nowy Sącz, (*37*) Ciężkowice, (*38*) Tarnawa, **Stowing sands**: (*39*) Upper Silesia, **Clayey raw materials and gravel aggregates**: (*40*) Jaroszów, (*41*) Tomaszów (dots signature—other smaller areas of extraction of clayey raw materials and gravel aggregates), abbreviations: (*CWC*) the Central West Carpathians, (*EBC*) the East Baltic Coast, (*EBL*) the East Baltic Lakeland, (*EEC*) the External East Carpathians, (*SCU*) the Silesian-Cracow Upland, (*SM*) the Sudety Mountains, (*SSL*) the Saxon-Lusatian Lowland

1993, 2008). Along with the process of the liquidation of mines, unemployed miners started to extract coal illegally from poverty shafts; in and around Sobięcin, about a thousand of these were accounted for. Some of these excavations collapsed and the resulting sinkholes reach a depth of up to 6 m (Wójcik 2006).

In the Lublin Coal Basin, one mine is in operation—the Bogdanka, located mostly in the area of the Lublin Polesie. Within the mining area (77 km^2), coal deposits are located almost horizontally, the deposit is tectonically slightly disturbed, and operations are currently carried out at a depth of 860–960 m (Taras et al. 2008). Since the establishment of the mine in 1983 until 2009 (that is, within 26 years), 79.3 million tonnes of coal were extracted, which is only 0.7 % of national production. However, from 2010 to 2014, production increased to as much as 31 million tonnes, and the Bogdanka mine has become the largest mine in Poland (7.7 million tonnes in 2014). The impact of mining on the relief was demonstrated in the form of surface subsidence; several troughs emerged, including two larger ones with water reservoirs with a total area of 1.6 km^2. A landfill with tailings also formed next to the mine (about 0.6 km^2), which, due to its height of 26 m, stands out in the flat landscape of the Polesie.

In Poland, 11.239 billion tonnes of coal were extracted until 2009, of which most is in the Upper Silesian Coal Basin (nearly 95 %).

Lignite is present in the Polish Lowlands and the Sudetes Mountains. Miocene deposits, whose industrial raw materials as of 2014 amounted to 1.2 billion tonnes, are of industrial importance (Bilans zasobów 2001–2015). Open-cast operation is carried out in three basins: the Turoszów Basin, the Bełchatów Basin, and the Konin Basin. Coal-bearing layers take the form of deposits or lenses, which occur in tectonic grabens and among glacitectonically disturbed sediments. Currently, operations are carried out in 9 pits within 4 mines.

The oldest lignite mining region is the Turoszów Basin in the Western Sudetes. From the initiation of operations in 1740 until the end of the nineteenth century, more than 100 small underground mines have operated here (Wilczyński 2010). In 1904, an open-pit mine, Turów I, was established; in 1963 came Turów II, currently with a depth of 210 m and over 24 km^2. The thickness of the exploited deposits is 50–60 m. Until 2009, 851 million tonnes of lignite were extracted from them.

One of the longest exploited lignite deposits in Poland is the Sieniawa deposit located in the Lubusz Land (Kozacki 1980). In 1873, an underground mine was established here, with a maximum production of up to 80,000 tonnes (1939). After rebuilding from the war destruction, the Sieniawa mine was active in the years 1950–1997. Since 2002, its place has been taken by an open-pit mine of the same name, yet with a low rate of production: 0.267 million tonnes in 2009.

The Konin Basin lignite deposits have a thickness of 8–10 m and are found under an overburden with a thickness of several dozen metres. Operation was initiated at the end of World War II and currently there are two active mines: the Konin and the Adamów. In the first one, extraction is conducted in 4 pits while 5 pits are inactive (Kozacki 1987). By 2009, more than 544 million tonnes of lignite had been extracted from the Konin mine; in the Adamów, more than 182 million tonnes (from 4 pits) have been extracted.

The youngest lignite basin is the Bełchatów Basin. The Bełchatów pit was established here in 1981 and is the largest in Poland (almost 26 km^2, and more than 200 m deep) (Fig. 8.2). The deposit occurs in the Kleszczowa tectonic graben at a depth of 150–200 m and is developed as a lignite deposit with an average thickness

Fig. 8.2 Bełchatów lignite open-pit mine in the Bełchatów High Plain, Poland (Dulias 2011)

of 50–55 m. Half of the Polish production of lignite came from the Bełchatów pit in the beginning of the 1990s; until 2009, 848 million tonnes of this raw material had been extracted from the pit. For several years, another pit—the Szczerców—has been in operation (almost 7 km^2, 120 m deep).

Since the beginning of lignite mining in Poland until 2009, over 2.4 billion tonnes of it had been extracted from 15 pits. The establishment of large-scale open pits was accompanied by the formation of large spoil tips of overburden. The movement of rock masses associated with the provision of the deposit reached a volume fourfold greater than the one associated with the extraction of the raw material. In the former Gosławice pit, 5.6 million m^3 of overburden were removed (Pilawska 1967); in the Bełchatów pit, it was as many as 3.6 billion m^3 (over 600 times more) (Tajduś and Kasztelewicz 2009).

The spoil tip with overburden from the Turów II mine covers an area of nearly 22 km^2, with a relative height of 245 m. Despite the large size of the form, as a result of the recently completed reclamation, it has blended into the mountainous landscape of the Żytawa Trough (Nietrzeba-Marcinonis 2007); on the other hand, the reclaimed spoil tip of the Bełchatów mine (referred to as Mount Kamieńsk, with nearly 15 km^2 and 1.35 billion m^3) rises 195 m above the undulating plain of the Bełchatów High Plain (Uberman and Ostrega 2004). Large quantities of overburden were removed from Konin mine excavations—so far, 2.8 billion m^3. In total, from the beginning of mining activity until the end of 2009, 9.5 billion m^3 of overburden was removed from Polish lignite mines, of which 38 % came from the Bełchatów mine (Kasztelewicz and Zajączkowski 2010). The (external) spoil tip formation outside open-pit mines comes to an end when parts of the deposits are depleted and the bottom of the excavation is made available for an internal spoil tip. Waste such as ashes, desulphurisation products, and municipal waste is also stored in the pits. Lignite mining operation is performed in an area covering between 200 and 250 km^2 (Maciejewska 2000).

Zinc and lead ores were mined in three districts—the Silesia and Cracow, Świętokrzyski, and the Lower Silesia. The flourishing of mining in the Świętokrzyski region happened in the fifteenth century and the first half of the sixteenth century; on every hill around Chęciny, there were several hundred shafts (Pazdur 1960). Subsequent attempts of restoring mining in this area have had little economic significance. In the region of Lower Silesia, lead and silver ores were extracted in the Świdnica-Jawor Land, in the area of Wałbrzych and Kłodzko. However, the region of main importance was, and still is, the Silesian-Cracow region. Operating areas located within the Upper Silesian Coal Basin are presented in Sect. 2.2. Outside the USCB, the most important and largest mining district was the Olkusz district. From the Middle Ages to the eighteenth century, ore mining here was comparable in volume to the production of other European centres; the landscape still features many post-mining forms from this period (Molenda 1972, 1978). In the early nineteenth century, there were 10 active open pit and underground mines, of which, after World War I, only two continued their operations—the Bolesław and the Ulysses. In 1931, as a result of the economic crisis, mines were immobilized for several years. After World War II, two large mines were established—the Olkusz (1968) and the Pomorzany (1974)—and the Bolesław mine was re-opened in 1945; until 1996, 28.5 million tonnes of ore had been extracted from it (Wójcik and Chmura 2005), whereas the other two mines had produced about 100 million tonnes. An area of approximately 33 km^2 was within the activity of underground mining. Currently, only the Pomorzany mine is open.

In Poland in the postwar period from 1945 to 2009, a total of 241.5 million tonnes of zinc and lead ores were extracted, of which about 110 million tonnes were in the area of the USCB and about 130 million tonnes were outside its borders in the Olkusz area. Shallow exploitation of zinc and lead ores contributed to surface deformations of the discontinuous type (Sroczyński 1997). Sinkholes associated with old, medieval exploitation are present in Olkusz Stary, for example, whereas sinkholes caused by modern operations occur mainly in the Baba valley and in the area of Hutki and Sztolnia Ponikowska. They are characterized by a regular cone shapes with a maximum diameter of more than 130 m and a depth of up to 40 m (Tyc 1989). During the most intensive mining period at the turn of the 1970s and the 1980s, almost 170 of them formed.

Technological processing of zinc and lead ores produces a large amount of flotation tailings—1.4 million tonnes in 2009 (Fig. 8.3). The amount of the previously accumulated waste in the Olkusz area was determined to be 38–50 million tonnes (Cabała 2005; Nowak 2008). Including the tailings of spoil tips in the USCB, between 150 and 200 million tonnes of flotation tailings accumulated in Poland in an area of approximately 3.5 km^2. The volume density of flotation

Fig. 8.3 Settling tank after flotation of zinc and lead ores in the Bolesław Mining and Metallurgical Plant (Dulias 2012)

tailings of zinc and lead ores is about 3 g/cm^3 (Trafas and Eckes 2007); hence, the volume of landfills may be estimated at 50–67 million m^3.

Copper ores were first mined in the Świętokrzyski region. In the 15th and 16th centuries, intense mining works, as for that time, were conducted in Miedzianka. The mine collapsed after running out of shallow deposits to the water level; the subsequent launches were short-lived, ineffective and ended in failure for various reasons, ultimately in 1920. In Lower Silesia, the construction of copper mines was initiated by the Germans in the 1930s; however, in practice, only after World War II was the exploitation of the rather poor deposits initiated in the area of Bolesławiec and Złotoryja in the Lena, Konrad, Nowy Kościół, and Lubiechów mines.

The Permian deposit in Lubin-Głogów, discovered in the 1950s and developed in the 1960s within the Fore-Sudetic Monocline, is at a depth of 800–1500 m with copper-bearing shales, sandstones, and dolomitic limestones. The thickness of the deposit ranges from 0.4 to 26 m, with an average of 3.4 m. Industrial resources have reached 1.2 billion tonnes (Bilans zasobów 2001–2015). The mining area is characterized by complex geological conditions of the overburden (highly hydrated layers prone to bounces). The initiation of mining in new mines—the Lubin (1967), Polkowice (1968), Rudna (1974), and Sieroszowice (1980)—provided a high position for Poland in the global copper mining industry (Smakowski et al. 1997). Ores are extracted using the chamber-pillar system with a deflection of roof rocks. A total of 1,700 km of active workings, with a capacity of over 30 million m^3, are maintained to date. The mining areas of these four mines occupy about 330 km^2, including 266 km^2 of deposits only. The area stretching from the Głogów ice-marginal valley in the north, through the Dalków Hills, to the Lubin Plain in the south, is within mining impact. In the postwar period in the years 1950–2009, a total of 993.3 million tonnes of copper ore were extracted, including only 60 million tonnes from the North Sudetic Trough (Kijewski and Downorowicz 1987).

The mining of copper ores is related to the production of large quantities of waste, generated almost exclusively during the flotation enrichment of ores (Traczyk 1997). In recent years, about 25 million tonnes of waste has been stored per year, which is more than the quantity of the extracted ore. Inactive landfill areas of about 9 km^2 have accumulated approximately 168 million tonnes of waste: 74 million tonnes in the Old Basin and 92 million tonnes in the Gilów landfill in the New Basin (Łuszczkiewicz 2000). Since 1977, flotation waste has been collected in the Żelazny Most settling tank; however, there have been discrepancies in the assessment of its weight and volume. About 10 years ago, the most frequently quoted volume was 350 million m^3. Since then, according to the National Geological Institute (Bilans zasobów 2001–2015), almost 250 million tonnes of new waste has been produced. Assuming that the volume density of copper ore flotation tailings is approximately 2.7 g/cm^3 (Bogda and Chodak 1995), it may be estimated that at present, the Żelazny Most accumulates waste with a volume exceeding 440 million m^3 in the area of 14.3 km^2 (38–40 m in height).

The consequence of the underground mining of copper comes also in the form of surface subsidence. The cities of Lubin and Polkowice, and several villages are located within the mining impact. In areas of intensive exploitation, the surface subsided by 2.0–2.6 m—up to 3.4 m in the area of the Polkowice mine. The surface lowering, due to exploitation overlaps with lowering related to the formation of a drainage trough, were, on average, 0.3 m, and a maximum of 0.71 m in the mining area of Lubin I (Krawczyk and Perski no date).

Rock salt mining dates back to the beginning of the Middle Ages and had been associated with Miocene resources in the Carpathian Mountains until the nighteenth century, when the Permian deposits in Kujawy became of interest. The oldest mine is the Bochnia, founded in 1248. In almost 750 years, 9 million tonnes of salt were extracted from the deposit of a complex geological structure, of several levels at depths from 70 to over 460 m. Within this mine, "dry" exploitation was carried out from the Siedlec-Moszczenica deposit (1975–1991) and from the Łężkowice deposit (1968–1988) using the borehole method (Garlicki 1999).

For about 700 years, mining was carried out on a much larger scale in the Wieliczka mine (now a UNESCO World Heritage Site), as evidenced by production reaching 25 million tonnes that created 2330 operating chambers connected by a 200-km system of galleries on 9 levels (Ślizowski and Saługa 1996). The Barycz deposit, exploited from the surface by leaching, also belongs to this mine (1924–1995).

Rock salt deposits in Kujawy are in the form of domes pushing through the Mesozoic cover at a depth of 100–300 m. Many such structures have been recognized, including some with heights of up to 7,000 m. In the Solno mine in Inowrocław, leach was extracted on an industrial scale from a dome to a depth of 180 m. In 1924, the construction of an underground mine (to a depth of over

650 m) was initiated. During almost 50 years, 100 million m^3 of brine was extracted and the volume of excavations amounted to 15.4 million m^3. Within the limits of the Wapno mine, mining initially started from the extraction of gypsum in 1826, which, at the beginning of the twentieth century, changed to the underground extraction of salt and gypsum at a depth of 38 m. The deep exploitation (400 m, and then almost 650 m) was initiated in 1920. Currently, the only mine in Poland that extracts rock salt in the traditional manner is the Kłodawa mine, built after World War II with a depth of 750 m. Until 2009, it had extracted 36.5 million tonnes of rock salt. Operations are also run in 3 mines that use the borehole method: the Góra (1200 m), Mogilno I, and Mogilno II. A relatively insignificant amount of salt is extracted from the Sieroszowice deposit in the overburden of the copper ore deposit. In the postwar period from 1945 to 2009, a total of 191.2 million tonnes of rock salt were extracted in Poland.

Salt mining has its consequences in surface deformations, both continuous and discontinuous. The latter were recorded in the impact zone of the Wieliczka mine as early as in the sixteenth-century historical sources. In modern times, the largest sinkholes (10 m deep) were created in 1977 within the Wapno mine limits after a disastrous inflow of water, eventually leading to the flooding of the mine (Langer 2007). Uncontrolled collapses occur in nearly half of the post-exploitation chambers in the Wieliczka mine. Surface deformations coupled with outflows of leach to the surface also accompany the processes of self-destruction of leaching chambers; for example, in 1974, a sinkhole with a depth of 8 m formed above the Barycz deposit and in 2001 above the Łętkowice deposit (9 m) (Traple and Wilk 2002). The phenomenon of suffosion also takes place.

Continuous deformations in the areas of the Wieliczka shafts have been controlled since 1926; they occur at a rate of 10–50 mm/year. For example, one of the troughs in the region of the main mine shaft decreased by a maximum of more than 1.5 m (Szewczyk 2008) in the years 1970–2005. The impact of underground mining in the Kłodawa mine is insignificant; resulting in two troughs of 0.8 and 2.8 km^2 and with a lowering of 0.08 and 0.14 cm, respectively (Ślizowski and Saługa 1996).

In Poland, **sulphur** occur in the Miocene deposits on the northern edge of the Carpathian Foredeep. Mining of native sulphur has a centuries-old tradition. From the fifteenth century to 1884, it was exploited from shallow but rich deposits in Swoszowice, in the Cracow region. A late-eighteenth century map of the mine reveals about 600 abandoned shafts associated with the period of the most intense activity of the plant. In the area of Cracow, sulphur deposits were also exploited in Posądza (1915–1921), gaining a total of about 300 tonnes (Preidl and Wójcik 2008). Sulphur mining also developed in the region of Kielce, in Czarkowe (with interruptions in the years 1795–1918), Czajków and Wola Wiśniowska near Staszów (Osmólski 1971). In the region of Upper Silesia, underground operations were conducted in the years 1878–1894 in Pszów and Kokoszyce. The total production of sulphur from the Middle Ages until 1921 is estimated at about 200,000 tonnes (Ney 2000).

The discovery of native sulphur deposits of resources on a global scale in 1953 was a breakthrough for sulphur mining; it was the largest in the Tarnobrzeg region (Piaseczno, Machów, Jeziórko, Osiek) and in the area of Staszów (Grzybów) and Lubaczów (Basznia). A series of deposits is composed of sulphur-bearing limestones, 8–20 m thick, occurring at a depth of 14–174 m. The sulphur content of the rock is, on average, 25–30 %, with a maximum value of 70 %. The deposits are in the form of decks and are poorly disturbed tectonically; they were excavated with the open-pit (Machów I, Piaseczno) and borehole methods (in other deposits). A total of 73 million m^3 of ores were excavated from the open pits, out of which 14.5 million tonnes of sulphur was achieved (Mika and Paszcza 2007). Together with underground mining, 126.6 million tonnes of sulphur were extracted in Poland in a period of 52 years, from 1958 to 2009. Most of the production came from the Jeziorko mine (74.1 million tonnes) and the Grzybów mine (26.4 million tonnes). The last active mine (both in Poland and in the world), Osiek, has produced an output of 10.4 million tonnes thus far. The area of mining impact covers almost 40 km^2.

Sulphur mining is related to the creation of anthropogenic relief forms in the shape of pits, overburden dumps, subsidence troughs in the areas of borehole operation, sinkholes, and sediment ponds. Excavation slopes built of Krakowiec clays have been undergoing landslide processes, whereas the pressure of landfills on the surface has caused the extrusion of ground in a distance of approximately 1 km. The volume of the Machów excavation (5.6 km^2) is estimated in the range of 336–370 million m^3 and the Piaseczno pit (1.6 km^2) at 77 million m^3. Both workings are reclaimed in the aqueous direction. The external landfills accumulated some of the overburden in the Machów landfill (8.8 km^2, 180 million m^3) and the Piaseczno landfill (1.2 km^2, 51 million m^3). In the areas of borehole operation, surface subsidence is 6 m in the area of the Jeziorko mine and 5–7 m in the area of the Osiek mine (Kowalik et al. 2009).

Gravel aggregates commonly occur mainly in Quaternary forms: glacial and glaciofluvial (in northern and central Poland), fluvial (mostly in southern Poland), as well as lacustrian, Aeolian, and marine. Extraction of aggregates is dispersed in several thousand locations across the country. For several years, it has had a dominant position both in the production of rock materials as well as all mineral resources mined in Poland (Fig. 8.4). Its extraction in 2009 amounted to 141 million tonnes. In the years 1960–2009 in Poland, over 3.8 billion tonnes of natural aggregates were extracted.

In Poland, there are large deposits of **stowing sands**, which are used to fill depleted underground workings; industrial resources amount to nearly 77 million m^3 (Bilans zasobów 2001–2015). The condition to account for a deposit in the resources is its location at a distance of not more than 50 km from the site of application. Documented deposits (Quaternary sand deposits and sand and gravel deposits) are in the region of the Upper Silesian Coal Basin (81 %) and in the Legnica-Głogów Copper District (19 %).

Until the mid-twentieth century, the annual production of stowing sands amounted from a few to a dozen million m^3; in subsequent years, it increased to 20–30 million m^3 (Wrona 1977). In 1970, it reached a record level of over 48 million

Fig. 8.4 Exploitation of the gravel aggregates in the Frydman mine, the Nowy Targ Basin (Dulias 2015)

m^3. At the end of the 1980s, due to the closure of many coal mines, the demand for sand decreased and its production gradually declined: it amounted to about 6 million m^3 in 2009 and 3.8 million m^3 in 2014 (Bilans zasobów 2001–2015). The most intense exploitation of sand was carried out on the outskirts of the USCB, which resulted in a significant transformation of the relief: a group of pits of the largest mine, Szczakowa, occupies an area of over 32 km^2. The total surface of the pits remaining after the mining of stowing sands in Poland was calculated and is partly estimated to be at least 95 km^2. The volume of post-stowing sand excavations totals about 1,300 million m^3, of which 80 % is within the USCB (excluding the sandpits of Kotlarnia, Pławniowice, Nakło-Chechło, and Hutki, which are situated in the neighbouring area but outside the USCB).

According to statistical data, the mining of stowing sands in Poland in the years 1960–2009 amounted to 1275.1 million m^3 (2,167.7 million tonnes). In the earlier period (the years 1900–1960), all of the sands were mined within the USCB, so their volume was already included in the calculation cited above (about 25 million m^3). It can be assumed that within 100–110 years, stowing sands mining transformed the area of about 95–100 km^2 by removing approximately 1,300 million m^3 of the mineral from the ground.

Dimensional and crushed stones have been extracted in Poland since the Middle Ages, such as Pinczów limestone, Strzelin granite, or Kunów sandstone. Currently, this group of rock minerals includes 33 lithological varieties of sedimentary (47 % of resources), magmatic (44 %), and metamorphic (9 %) rocks. Of the 685 deposits, 292 are permanently or temporarily extracted, while extraction was abandoned in 155 of the deposits. Industrial resources of the described minerals are reported at almost 3.6 billion tonnes (Bilans zasobów 2001–2015). Based on data from the National Geological Institute (the MIDAS database), Nita (2010) calculated that there are 730 medium and large quarries in Poland, with over 32 % of them being sandstone quarries and almost 24 % limestone quarries. Quarries of metamorphic and magmatic rocks are concentrated primarily in the Sudetes and the Sudetic Foreland and in the southern part of the Cracow Upland, while the exploitation of sedimentary rocks—limestone, dolomite, and sandstone—is conducted in the Carpathians, in the Świętokrzyskie Mountains, across the Polish Uplands, in the Sudetes, and in the Sudetic Foreland.

In 2009, 55.3 million tonnes of rocks were extracted, mostly dolomites (17 %), basalts (15 %), granites (13 %), limestones (12 %), and sandstones (10 %). In that year, the biggest production was achieved by the migmatites quarry in Upper Piława (2.7 million tonnes), the basalts quarry from the Krzeniów deposit, the melaphyres quarry from the Grzędy deposit (over 1.8 million tonnes each), the dolomite quarry from the Nowa Wioska deposit, and the gabbros quarry from the Braszowice deposit (1.7 million tonnes each). From 1990 to 2009, a total of 508 million tonnes of dimensional and crushed stone was extracted in Poland.

Limestones and marls for the cement and lime industries occur primarily in the regions of Opole, Cracow-Wieluń, Świętokrzyski, and Lower Silesia. In a majority (60 %), they are Jurassic deposits. In 2009, however, the biggest production came from the only deposit in northern Poland, Barcin-Piechcin in the Kujawsko-Pomorskie (4.8 million tonnes) and from the deposits in Ostrówek and Ołowianka in Świętokrzyskie (3.4 million tonnes; Bilans zasobów 2001–2015). Industrial resources of limestones and marls amount to 3.2 billion tonnes. In the last 50 years (1960–2009) a total of over 1.9 billion tonnes of limestones and marls were excavated for the cement and lime industries (Fig. 8.5). Out of the 186 deposits, 43 have discontinued their operations and 34 are still open for mining.

Dolomites with the best quality parameters are applicable in metallurgy and should be accounted for independently from dolomites used as building stone and crushed aggregates. The deposits are Devonian or Triassic. They are mined from the deposits of Bobrowniki-Blachówka and Gródek and continue to be extracted from the deposits in Brudzowice-Siewierz, Ząbkowice Będzińskie, and Żelatowa (Fig. 8.6). From a record production volume of 5 million tonnes in 1979, it has dropped to 2.3–3.7 million tonnes per annum in recent years. Industrial resources amount to 71.8 million tonnes, of which half is located in the Brudzowice deposit. In the years 1960–2009, a total of 144.3 million tonnes of dolomites were extracted for the steel industry.

Clay materials are common materials exploited since the mid-thirteenth century. Large companies, however, began to emerge in the second half of the nineteenth century. The clays that are mainly extracted include ice-dammed Quaternary clays, Neogene Poznań clays, and marine clays of the Carpatian Foredeep (about 76 % of the total extraction). Triassic and Jurassic clay exploitation is of lesser importance. The number of clay material mines has been steadily decreasing, but

Fig. 8.5 Limestone quarry in the area of the Wierzbica mine, Sobków in the Nida Basin (Dulias 2004)

Fig. 8.6 Dolomite quarry in Ząbkowice Śląskie, the Ząbkowice Hummock (Dulias 2011)

Fig. 8.7 Clay pit in Busko in the Nida Basin (Dulias 2004)

there are still hundreds of them scattered throughout Poland (Galas 2004) (Fig. 8.7). The total production of clays has decreased from over 14 million tonnes per year in the 1970s to 5–6 million tonnes/year in the last decade. The largest mining is carried out from deposits in Chwalimierz (the Lower Silesian Voivodeship), Lębork (the Pomeranian Voivodeship), and Tadeuszów (the Świętokrzyskie Voivodeship). In the years 1960–2009, a total of 507.8 million tonnes of clay materials were extracted in Poland.

As a result of the dynamic development of mining in Poland after World War II, the old mining regions of energy resources, ores, and chemicals developed and new ones were established. Currently, of these 12 districts, some have lost their purpose; for example, the Częstochowa Iron Ore District, the Tarnobrzeg Sulphur District, and others play a significantly lesser role in the national economy, such as the Silesian-Cracow District of Zinc and Lead Ores. The plans for establishing 18 raw materials exploitation regions have not been realized, despite their approval by the Council of Ministers in 1978 (Kozłowski 1983). Nevertheless, it can be assumed that the largest mining rock mass movement in Poland took place or continues to take place within 30 mining areas. They are concentrated in the southern part of the country from the Carpathians and the Sudetes, through the Polish Uplands to the Wielkopolska Lowland. In the north, and particularly in the north-eastern part of the country, mining plays a subordinate role.

The scale of the mining rock mass movement is illustrated by the data compiled on the basis of the balance of resources regarding the number of active and inactive open-pit mines in Poland in 2009: 4,355 excavations of natural aggregates, 673 quarries of solid rocks, 917 excavations of clay materials for construction ceramics, and 283 pits of other rock materials (Bilans zasobów 2001–2015). Regardless of these 6,228 locations of rock mining, there are countless smaller pits in which operations were carried out and are still carried out on a local scale. Such a significant "fragmentation" of mining production makes it difficult or virtually impossible to collect data about the morphometric characteristics of the workings (their size, depth, volume, and duration of activity) and a more detailed analysis of the volume and intensity of movement of rock masses. In 2009, 34 traditional underground mines were active as well: 28 coal mines, 4 copper mines, 1 salt mine, and 1 lead and zinc mine. There were 4 mines that used the borehole method, including 3 rock salt mines and 1 sulphur mine.

A summary of the extraction of mineral resources in the last 50 years (1960–2009) shows that it amounted to more than 20.6 billion tonnes. The exploitation of rock materials (47.3 %) and energy resources (45.4 %) were of main importance. The share of other resource groups was relatively low: metallic materials (5.8 %) and chemicals (1.5 %). Among individual raw materials, the highest share in the total extraction was given to coal (33.9 %), followed by natural aggregates (18.6 %), lignite (11.5 %), and stowing sands (10.5 %). These four minerals accounted for almost 75 % of production in the reported period. Geomorphological units with the largest displacement of rock masses include the Bełchatów Upland in central Poland and the Silesia Upland in the south of the country (within the USCB).

In the years 1960–2009, a total of 41 % of the extraction of mineral resources in Poland came from underground operations conducted in an area of about 2,500 km^2. On average, 3.4 million tonnes of materials were extracted under every 1 km^2 of the mining area. The area occupied by open-pit mining is difficult to be precisely determined because excavations are included in the statistics, together with other land occupied by mining activities such as landfills or areas of industrial development. According to estimates, the area of open-pit mines covers at least 300 km^2; therefore, in the past 50 years, an average of 40.5 million tonnes of minerals were extracted from each 1 km^2, which makes it 12 times more than in the case of underground mining. The more intense movement of rock masses is therefore associated with open-pit mining. The annual loss of rock masses from the bedrock in the case of underground mining amounts to 68,000 tonnes/km^2, whereas it is 810,000 tonnes /km^2 in the case of open-pit mining.

The exploitation and enrichment of minerals, regardless of the method of operation, is connected to the mining, processing, and storage of waste, as well as the dumping of overburden from open-pit mines. In 2009, the volume of spoil tips amounted to at least 10.8 billion m^3 (Table 8.2). In the case of some materials, such as copper ores, the amount of waste produced annually exceeds the amount of the extracted mineral; similarly, the amount of the removed overburden in lignite mines exceeds its production. In the structure of the dumped material, major importance is given to the overburden from lignite mines: it represents 88.3 % of all rock masses accumulated on spoil tips in Poland. In statistical studies, the overburden is not recognized in the balance of waste. According to the Central Statistical Office data, in the years 1980–2009, the amount of waste that accumulated in Poland varied in

Table 8.2 Mining wastes and overburden landfilled in Poland until 2009 (from Dulias 2013)

Raw materials	Mining wastes and overburden			
	(mln m³)	(%)	Area (km²)	Average height (m)
Coal	478	4.4	53.6	8.9
Lignite	9,546	88.3	60.0	159.1
Copper ores	502	4.6	23.3	21.5
Zinc and lead ores	67	0.6	3.7	18.1
Sulphur	231	2.1	10.0	23.1
Total	10,824	100.0	150.6	72.8

the range from 0.9 to 2 billion tonnes. In 2009, it amounted to 1.74 billion tonnes. According to the Institute of Waste Management in Katowice, on Polish spoil tips, there may be as much as 4 billion tonnes of waste. It is most often assumed that waste associated with the mining industry accounts for about 80 % of the total amount of waste. Hence, its number would be 1.4–3.2 billion tonnes and the estimated volume would be 0.5–1.2 billion m^3. The cubic capacity of mining waste (excluding overburden and rock mining waste) is 1.05 billion m^3 (Dulias 2013).

On the basis of the data presented in this chapter, it may be concluded that in the last half-century (the years 1960–2009), the share of the Upper Silesian Coal Basin in the total mining of mineral resources in the country was quite large, amounting to as much as 45 % (more than 9.2 billion tonnes). No other area in Poland was in the range of such intense mining activity. The amount of accumulated waste against the background of the whole country is insignificant: it is 3.4 %, but it occupies 33 % of the total of spoil tips in the area in Poland.

Denudation. The volume of rock mass movement associated with mining in Poland in 1960–2009 became the basis of calculations and estimates of the size of denudation and anthropogenic aggradation in selected areas. Convex and concave anthropogenic forms associated with open-pit and underground mining of various mineral resources were examined and reliable data were obtained.

The anthropogenic aggradation rates A_A calculated for spoil tips reached very high values—from a few hundred to a few thousand mm/year and up to 5,366 mm/year in the case of the Bełchatów external spoil tip (Table 8.3). The rate of anthropogenic aggradation A_A in underground mining areas is smaller but quite varied. For example, the Victoria coal mine spoil tip's anthropogenic aggradation (9/2) in the Lower Silesian Coal Basin, based on data included in Wójcik's work (2006), was calculated at 372 mm/year; the average for all spoil tips in the Wałbrzych region is 183 mm/year. In the USCB, the A_A indicator amounts to 175 mm/year for the Marcel mine spoil tip and 1,258 mm/year for the Krupiński mine spoil tip.

The intensity of anthropogenic denudation is generally higher in areas of open-pit mining than underground mining. The D_A indicator in the area of the Piaseczno openpit sulphur mine is 3,702 mm/year, which makes it dozens of times higher than the rates of anthropogenic denudation in areas of underground mining

Table 8.3 Anthropogenic denudation and aggradation rates for the selected mining areas in Poland (after Dulias 2013)

Anthropogenic denudation rates D_A (mm/year)		Anthropogenic aggradation rates A_A (mm/year)	
Opencast mining			
Machów sulphur open mine	2627	Spoil tip of the Machów sulphur open mine	818
Piaseczno sulphur open mine	3702	Spoil tip of the Piaseczno sulphur open mine	3269
„Bełchatów" lignite open mine	3000	Spoil tip of the Bełchatów lignite open mine	5366
„Adamów" lignite open mine	404	Spoil tip of the Bełchatów lignite open mine (Szczerców field)	2868
„Koźmin" lignite open mine	729	Spoil tip of the Turów lignite open mine	1142
„Pątnów" lignite open mine	675		
Lubstów lignite open mine	1898		
„Pogoria III" sandpit	574		
„Żelatowa" dolomite quarry	326		
Żychcice limestone quarry	105		
Underground mining			
Subsidence troughs in the Lower Silesian Coal Basin (Wałbrzych field)	53	Spoil tips in the Lower Silesian Coal Basin (Wałbrzych field)	183
„Rudna"copper mine	23	„Żelazny Most" settling tanks in the Lubin-Głogów Copper District	933
„Polkowice" copper mine	15	Spoil tip of the „Krupiński" coal mine	1258
Jeziorko sulphur mine	41	Spoil tip of the „Moszczenica" coal mine	235
Grzybów sulphur mine	49	Spoil tip of the „Borynia" coal mine	822
„Kłodawa" rock salt mine	14	Spoil tip of the „Ziemowit" coal mine	354
„Wieliczka" rock salt mine (Barycz field)	54	Spoil tip of the „Pniówek" coal mine	171
"Olkusz-Pomorzany" zinc and lead ores mine	19	Spoil tip of the „Marcel" coal mine	175
"Orzeł Biały" zinc and lead ores mine	10	Spoil tip of the „Dębieńsko" coal mine	80

of various mineral resources; for example, in the area of the underground exploitation of rock salt in the Kłodawa mine, it is 14 mm/year. On the Cracow-Wieluń Upland, open-pit exploitation of rock materials resulted in a local lowering of its surface by an average of 18 mm, up to 4,039 mm (Szczypek and Trembaczowski 1987); after appropriate conversion, this gives an average rate of anthropogenic denudation of 0.8 mm/year, with a maximum of 202 mm/year. The author's calculations indicate that anthropogenic denudation rates vary depending on the type of mined rock: it is significantly lower for hard rock quarries and greater for loose, crumbly rock. For example, for the Rogoźnik limestone quarry on the Twardowice Plateau, the anthropogenic denudation was estimated at 52 mm/year, whereas the Szczakowa sand pit in the Biskupi Bór Basin is at 359 mm/year. Based on the above data, it can be concluded that denudation, as well as anthropogenic aggradation, occur at a higher rate in the areas of open-pit mining. The spatial range of the impact of these processes, however, is far greater in the case of underground mining—more than eightfold (300 and 2,500 km^2 respectively).

Denudation and anthropogenic degradation vary over time. In open-pit mining, there are three phases of transforming the environment—the preliminary (dynamic changes in a short time), the actual (permanent changes, but less intense, often in leaps), and the reclamation phase (Radwanek-Bąk 2001). In the case of open-pit mining of sulphur and lignite, the initial phase of making the deposit available is associated with removing huge amounts of rock overburden. The D_A denudation rates within the open pit and the A_A aggradation in the area of overburden dumping are similar (theoretically equivalent) and clearly dependent on the thickness of the overburden. They reach very high values, from a few hundred to a few thousand mm/year. Rock material exploitation (natural aggregates, stowing sands, or hard rock) usually does not require removal of such a large amount of overburden. Therefore, the lowering of the surface is primarily associated with the main phase (the deposit extraction), and the elevation of the surface (not necessarily outside the excavation) is associated with accumulation of tailings.

Movement of rock material related to the reclamation of post-open pit mining landforms begins in the main phase on the fields that have been exhausted. The scope of earthworks depends on the direction of reclamation, but it generally includes the formation of excavation and spoil tip slopes and the levelling of the bottom of the pits or the heap tops. A fairly common practice is also a partial filling of mining voids with waste from other sources and the removal of overburden from neighbouring pits. For example, the Bogdałów lignite pit was filled in with overburden removed from the Koźmin open pit; reclamation plans for the Szczerców open pit include making it more shallow with the use of overburden from its own external dump (Wachowiak and Wachowiak 2004). The scale of rock material movement in the reclamation phase may be fairly large; in the Maczki Bór Zachód sand pit located in the Silesian Upland, more than 200 million tonnes of mining waste have already been accumulated. As a result of such actions, geomorphological effects of denudation and anthropogenic aggradation from the initial and main phases will be reduced or nullified.

Underground mining—which in Poland concerns coal, rock salt, sulphur, and the ores of copper, zinc, and lead—has a different specificity of rock mass movement than open-cast mining, which affects the volume of anthropogenic denudation. In the case of surface exploitation, an excavation of the size of the removed minerals, overburden, and waste rock is formed as a result of deep underground operations. The excavation is formed under the surface and to different degrees; depending on many factors, it might be revealed in the form of a subsidence trough or a sinkhole. The volume of the underground excavation is not equal to the volume of the form "reproduced" on the surface. In extreme cases, the underground excavation may never be reflected on the relief.

Anthropogenic denudation associated with underground mining, as in the case of open-pit mining, varies over time. During operations carried out at great depths, the denudation rate is dependent on the progress of the exploitation front, wherein the lowering of individual points of the surface does not end after the front but still lasts for an extended period of time (Skinderowicz 1974). The process of surface lowering typically ceases within 2–3 years. In areas where shallow operations were carried out, to a depth of 80–100 m, surface lowering may occur even after 100 years. Therefore, the period of anthropogenic denudation does not necessarily coincide with the period of mining activity, but may appear as its "echo" remote in time.

In Polish geographical literature, data on the size of anthropogenic denudation in the areas of underground mining relates primarily to the Silesian-Cracow Upland. The role of subsidence processes caused by the mining of zinc and lead ores in the contemporary denudation system of the Olkusz district was estimated by Tyc (1997) based on the volume of rock material displaced across all inventoried post-mining anthropogenic landforms. For the area covered with discontinuous deformations, the volume of 0.99 million cubic metres was achieved within 5 years of their intensive formation (1978–1983). The intensity of anthropogenic denudation in the whole area of hydraulic depression was calculated by this author at 0.6 mm/year, with the provision that, in reality, the process is limited to small areas; therefore, the achieved volume is actually greater by one order of magnitude—60 mm/year. It has been concluded that the intensity of anthropogenic denudation in the research area is almost 10 times greater than the chemical denudation measured by the outflow of dissolved substances in a year.

For areas of coal mining, the following anthropogenic denudation volumes are quoted: 71 mm/year for the study area within the Borynia mine (Madowicz 2001), 46 mm/year for the area of 8 mines located on the Murcki Plateau (Kupka et al. 2005), and 39 mm/year for mining areas in the eastern part of the Miechowice Upland (Solarski and Pradela 2010). On the basis of data contained in the work of Aleshina et al. (2008), it was calculated that the D_A indicator for the research area within the Pniówek and Zofiówka mines is 41 mm/year; for the Knurów and Szczygłowice mines, based on data by Wojciechowski (2007), it was 29 mm/year. These values are generally consistent with the ones presented herein.

Similar calculations for the D_A indicators were made for copper and sulphur mining areas. On the basis of the volume of the extracted copper ore, with the assumption of a roof operation coefficient with roof-collapse extraction at 0.45 (Strzałkowski 2010), it was estimated that the lowering of the mining area in the Legnica-Głogów Copper District (LGCD) was at an average of 12 mm/year; for the area of the mining impact, these values would be much greater. These theoretical calculations are generally consistent with data quoted by Krawczyk and Perski (no date), which shows that the maximum speed of the increase in surface lowering in the LGCD did not exceed 30 mm/year. On the other hand, in areas of the underground (borehole) sulphur exploitation, the surface lowered on average by 1–1.8 m within the limits of the Jeziórko mine, 1.3–1.5 m in the area of the Grzybów mine, and 0.9–2.9 m in the case of the Osiek mine; these are estimates based on data contained in works by Kowalik et al. (2009). Given the above, the D_A anthropogenic denudation indicator was estimated at 23–41, 41–49, and 54–171 mm/year, respectively.

Anthropogenic relief forms associated with both open-pit and underground mining are modelled by geomorphological processes. Lignite and sulphur pit slopes are or have been within landslide processes, which are fostered not only by their large inclination, but also clay and ductile overburden and ground vibrations caused by the use of mining equipment. Quarry walls are relatively the least exposed to denudation processes, which does not mean that these processes do not occur. The slopes of the spoil tip of the Turów II mine were especially prone to the processes of downcreeping, mass movement, and water erosion, especially in the initial period of its creation (Pulinowa 1967). According to Pilawska (1967), the leaching of the spoil tip material in Turów, built of about 70 % of kaolin clays, was then 3.7 m^3/km^2. Repelewska-Pękalowa (1973), on the basis of 5-year field measurements, calculated the intensity of flushing in the spoil tips of the sulphur mine in Piaseczno at 24 kg/m^2/year. Wójcik (1995), when describing mass movements on heaps in the Wałbrzych area (composed primarily of clay shales, claystone, and siltstone susceptible to rainwater) stressed that landslide tongues reach even several meters beyond the outline of the heap.

Post-mining and mining anthropogenic forms are also subject to aeolian processes, in particular with regard to the bottoms of excavations after the exploitation of sand. One of the most studied areas in this regard is the stowing sand mining area in the eastern part of the Silesian Upland (Szczypek and Wach 1991a, b; 1993). The measure of the intensity of aeolian processes occurring there may be the mass of sand removed from one of the deflation surfaces—240 kg/m^2/year. Such studies were also conducted in the Bełchatów pit, where the mass of sand moved during a sand and dust storm was estimated between 57 and 164 kg/m^2/h, while the annual precipitation of dust in the bottom of the mine was between 128 and 262 g/m^2 (Goździk et al. 2009). Aeolian processes also include fine-particle, post-flotation tailings from mining copper and zinc and lead ores, which pose a serious problem in the reclamation of settling tanks (Mizera 1980; Szczypek and Wach 1991b; Mizera and Nierzewska 1996; Trafas and Eckes 2007).

Other processes that take place in mining areas include the natural movement of rock masses caused by the prolonged and intensive drainage of deposits, which is induced by human activities. At the beginning of operations of the Bełchatów mine, the dynamics of subsidence, as a result of dehydration, was even 0.3 mm/day (Stawiarski and Malinowski 1983). An average volume of subsidence caused by dehydration in the sulphur mining area was 7–11 mm/year (Krawczyk and Perski no date).

Mining areas in Poland are located within neotectonic movements, both lifting and lowering. The "separation" of neotectonic movements from technogenic movements is not always possible. In Upper Silesia, the Vistula river basin decreases from −0.5 to −1.5 mm/year; the Oder river basin decreases from −1.0 to −3.0 mm/year (Kowalczyk 1964). In the Lower Silesia, extreme values of vertical movements of the Earth's crust, identified on the basis of interpretation of hydrogeological data, range from—2.14 to +1.55 mm/year (Badura and Wojtkowiak 1983). In general, it may be assumed that the calculation of anthropogenic denudation of mining areas located in tectonically active zones should take into account an error of ± 1 mm/year.

8.2 The Ostrava-Karvina Coal Basin, Czech Republic

The Czech part of the Upper Silesian Coal Basin covers an area of 1.550 km^2, including the Ostrava-Karvina Coal Basin (OKCB) at 320 km^2 (a few percent of the total area of the Upper Silesian Coal Basin, Polish and Czech) (Dopita 1997). The spatial extent of the basin is determined by the Polish-Czech border in the north, the foothills of the Beskidy Mountains in the east, and the Oder valley in the west. Generally speaking, the basin stretches between the cities of Ostrava in the west and Karvina in the east; towards the south, it reaches Frenštát.

In terms of geological structure, the area of the Ostrava-Karvina Coal Basin belongs to the Western Carpathians and the Bohemian Massif. The substrate is built of Paleozoic Devonian and Carboniferous sediments. In the relief of the Carboniferous roof, the highlighting forms are the generally latitudinally arranged ridges (Příbor-Cieszyn and Ostrava-Karvina) and lowerings (dětmarovicki vymol and bludovicki vymol). The overburden mainly consists of silty-clay sediments of the Lower Baden (Dopita 1997; Martinec et al. 2005). They have a varied thickness: in these depressions, it may reach 1,400 m; in areas where the Pleistocene glacial erosion removed part of the Miocene deposits, it is usually 250 m. In some places, directly located on the Carboniferous substrate, glacial and loess sediments are present.

Demek (1987) includes the northern, main part of the basin to the Ostrava Basin and the Moravian Gate. The greater part of the mining area belongs to the Ostrava Valley, and mainly to one of its units—the Orlovski Plateau (Orlovska plošina). The Carboniferous substrate is covered here by glacial sediments (sands, gravel, and clay). On the surface, there are loess and loess-like forms—remnants of older glacial moraines. The surface of the plateau is dissected by ravines, with frequent landslides. The plateaus located in the east and south of the Olrovski; the Karviński and Havířovski are similar. Plateaus with hilltops are at altitudes of 250–300 m above sea level, lower towards the wide valleys that surround them—to the west, towards the Oder valley with Ostravica, and to the north, towards the Olza valley (210–220 m above sea level). The southern part of the basin belongs to the West-Beskidian Piedmont—more specifically to the flysch Podbeskydská pahorkatina, rising to 300–600 m above sea level (Kirchner and Hradek 2004).

In the pre-mining period, the basin area was sparsely inhabited and agricultural. In the last 200 years, it was transformed into a densely populated mining and industry region. The beginnings of mining were associated with the discovery of coal on the Landek hill in Ostrava in 1763. The first mine was established in 1776, but data on production volumes started to appear from 1782 and are incomplete

(Martinec and Schejbalova 2004). Among the oldest mines located here are Frantisek, Anselm, Zofie, Michał, and Bettina.

The total amount of the coal mined so far is difficult to determine, especially for the pre-war period. Mining, which was initially insignificant, increased steadily with the growing demand for coal, particularly by the Vitkovice ironworks (1829), then by the railway. This is illustrated by coal production in the following years: 6,000 tonnes in 1822, 60,000 tonnes from 1842, 610,000 tonnes in 1862, and 1.2 million tonnes in 1872. In the early twentieth century, the extraction in the basin reached 6 million tonnes; in 1930, it was 10 million tonnes; and in 1943, it was 20 million tonnes (Martinec and Schejbalova 2004). It is estimated that by 1946, a total of 522 million tonnes of coal had been extracted in the basin, of which 61 million tonnes was until 1900.

In the postwar period, there were several active mines, including Julius Fučík, Jan Šverma, Hermanice, Ostrava, Dukla, and ČSA. New mines were also established: the 9. květen, ČSM, Paskov, and Starič in the southern part of the basin. The youngest mine, Frenštát, was established in 1981. Annual production over about 3 decades amounted to over 20 million tonnes. Operations were mainly carried out by longwall extraction with roof-collapse. As a result of mining restructuring, in the years 1992–2000 all mines in the region of Ostrava and Petřvald were closed, along with the František mine in the Karvina area and the Paskov mine in the southern part of the basin. In the period 1947–2000, nearly 1,089 million tonnes of coal were extracted (Schejbalova 2003). This represents more than two-thirds of the total coal production in the period from 1782 to 2000, which was 1,611 billion tonnes. According to Dopita (1997), total coal production in the years 1782–1996 was lower at 1,419 billion tonnes. Statistics compiled by Dulias (2013) show that coal production until 2005 was over 1.9 billion tonnes, which seems to be an overestimate. In 2000, the total mining area (active and inactive) amounted to 320 km^2. In 2009, there were 5 active mines with a total annual extraction of about a dozen million tonnes (in 2001: 15.2 million, in 2008: 12.6 million). Currently, 3 mines operate on the area of almost 134 km^2.

Particular parts of the basin vary in terms of the volume of extraction. In the last century (1901–2000) in Ostrava, which is the largest in terms of area (157.4 km^2), exploitation amounted to 635 million tonnes of coal. In the much smaller Karvina part (99.5 km^2), it was as much as 865 million tonnes. Coal production in the southern part (63.1 km^2) was 50 million tonnes (Martinec et al. 2005). The mines with the largest extraction in the analysed period included the Julius Fučík (202.5 million tonnes), the Ostrava (191.2 million tonnes), the Odra (159.9 million tonnes), and the Darkov (154.2 million tonnes).

The owner of almost all post-mining land is the private company OKD a.s. (Ostravsko-karvinské doly), which deals with the exploitation of coal and the reclamation of the post-mining land, conducted by the specially established IMGE company. In 1999, a list of degraded land was initiated in the form of digital maps and descriptions of sites. The database already covered 90 % of such areas. The former mining areas that were created before the privatization of the coal industry since 2002 have been managed by the state enterprise DIAMO s.p. (Santorius et al. 2007).

Hard coal mining in the Ostrava-Karvina Basin has resulted in significant relief transformations, and in some areas the creation of typical anthropogenic relief with a dominant share of forms resulting from direct and indirect human activities. Anthropogenic landforms mostly consist of subsidence troughs, spoil tips, and settling tanks, whereas discontinuous deformations are generally poorly visible in the morphology of the terrain. Discontinuous deformations were created and, from time to time, are also formed today, mostly in the areas of older mining on Carboniferous outcrops. They are mostly funnel-shaped forms. Thresholds appeared particularly frequently during shallow operations of thick coal seams. Sinkholes are also formed as a result of the collapse of shafts. The biggest destruction of a shaft in the history of the Czech coal industry was the collapse of the Doubrawa IV shaft, which took place in 1999. The shaft, with a depth of 1,100 m and a diameter of 8.5 m, was destroyed along its entire length. Within 24 h, a 36-m-deep crater formed on the surface with a diameter of more than 60 m and a volume of 65,000 m^3 (Aldorf et al. 2000).

In the area of deep exploitation (200–1,200 m), a number of subsidence troughs formed, which are among the most extensive anthropogenic forms in the basin. Already in 1965, they covered an area of 130 km^2, and some of them reached a depth of over 20 m. The depressions of up to 1 m had the biggest range, representing 80 % of the total area of mining impact. The subsidence areas over 5 m occupy only 3 % of the area, but they contributed to serious damages to the environment and land development (Mareš 1975). Due to extraction from the protective pillars, the town of Poruba was destroyed; in its place, the town of Havirov was built, with all necessary protections. Also, the town of Katarzyna, which belonged to Stonava, was torn down and further covered with waste rock and reclaimed. Approximately 70 % of the population of Olrova was evicted and the town of Karvina Doly disappeared from the map, leaving behind only the most famous example of mining damages—the church of St. Peter from Alcantara, tilted due to subsidence of the protection pillar slope of the Doubrava mine (Fig. 8.8). Since 1950, the surface in this area has lowered by about 34 m (Střítežská and Rafajová 2004).

The volume of subsidence in the basin depends on many factors– not only geotechnical, but also on the thickness of the overburden or the Carboniferous roof relief. The scale and method of exploitation of minerals, though, have a major impact on the formation of troughs. In the Ostrava-Karvina Basin, the development of subsidence after 1960 was intense for the following three decades: in the years 1961–1989, according to analysis by Hortvik (2003), 36.8 % of the impact area (251.3 km^2) lowered by more than 1 m and 63.2 % lowered by less than 1 m; in the decade from 1990 to 1999, the numbers were 15.1 and 84.9 %, respectively, but the mining impact covered a significantly smaller area at 179.4 km^2.

According to the OKD, the range of subsidence in 2000 (estimated according to the Budryk-Knothe method and designated by subsidence isoline 0 m) covered an area of nearly 254.8 km^2 (80 % of the basin), which was almost two times more than in the 1960s. In the Ostrava area, 131 km^2 were within the subsidence range; in the Karvina area, it was 79.3 km^2; while in the southern part, it was 44.5 km^2.

Fig. 8.8 Church of St. Peter from Alcantara in the deep subsidence trough (34 m) in the area of destroyed village of Karvina Doly in the Ostrava-Karvina Coal Basin (Dulias 2012)

The subsidence volumes are spatially varied. Even though 51 % of the mining impact area is located in the Ostrava area, it is characterized by the lowest surface subsidence, at an average of 4 m (maximum 10 m). This is due to two reasons: operating restrictions due to protective pillars and the extraction of thin coal seams (1.2 m) from the thicker carbonless rock layers (Schejbalova 2003). An average surface lowering for individual mines is varied: for example, for the Hermanice mine, it is 2.6 m; for the Ostrava mine, it is 4.7 m. In the Karvina area, where the exploited deposits are thick (an average of 3.4 m in thickness) and located between thicker layers of sandstone, the volume of subsidence is larger. The depths of the larger troughs here may reach between 15 and 20 m and maximally to 50 m (Martinec and Schejbalova 2004), which seems to be an over-evaluation. The mining area of the Karvina zone has thus lowered to a much greater extent than in the Ostrava zone—an average of 7.6 m. An average lowering in the area of the Lazy mine is as high as 13.5 m, the "Darkov" mine is 10.3 m, and the ČSA mine is 9 m. In the Karvina zone, a specific landscape has developed. On one hand, large subsidence areas have formed (land-locked or with partial natural drainage, partly used as settling tanks). On the other hand, the remaining elements are hills in place of the former protective pillars of shafts, as well as the waste collected on heaps. The original relief of a glacial plain has been transformed into a typically anthropogenic relief (Martinec and Schejbalova 2004).

In the Ostrava part of the basin, where operations were completed in 1994, surface movements have decreased or ceased; in the Karvina part, the volume of subsidence may increase due to operations reaching the marginal part of Carboniferous deposits and increasing its depth. In the years 1961–1999, the volume of subsidence in the basin amounted to 0.434 km^3, resulting from the creation of a void in the bedrock with a volume of 0.627 km^3 (Hortvik 2003). In the decade of 1990–1999, the subsidence volume amounted to 0.085 km^3 and reflected the void in the bedrock with a volume of 0.119 km^3. An average surface subsidence calculated on the basis of the above data in the years 1961–1989 was about 1 m; in 1990–1999, it was about 0.5 m. The forecasts of mining impact until 2010 assumed that the largest subsidence, exceeding 10 m, would occur in Darkov on the west side of the Olza River (Dopita 1997). Field observations confirm these predictions (Fig. 8.8). Martinec et al. (2005) emphasized, however, that there are significant differences between the predicted and the measured values of subsidence and used the example of the Starič mine, where the measured surface subsidence was half of the forecast.

Flood plains forming in subsidence troughs are mostly filled in with mining waste, such as in the village of Kunčičky at the confluence of the Ostravica and the Lučina in the mining field of the "ČSM" mine. Many troughs are used as settling tanks, such as in the area of the "Horni sucha" mine. Some water reservoirs in the troughs are used for recreation (Martinec et al. 2005).

The development of subsidence troughs is a factor destabilizing the balance of the slope. In the area of the basin in places inherently susceptible to mass movement, it was observed that some landslides are partially linked to mining activity. The main importance is the creation of discontinuous deformations in the form of fissures in which water infiltrates faster and deeper to the surface of rupture. The impact of mining activities on the activation of the subsidence process has been proven on the left bank of the Olza around Doubrava, as well as in Karvina Doly (Müllerova and Ides 2004).

Spoil tips and settling tanks are characteristic elements of the anthropogenic relief of the Ostrava-Karvina Basin (Fig. 8.9). The highest elevation in the Ostrava part of the basin is a heap located within the former Petr Bezruč mine, rising 325 m above sea level (Dopita 1997). An example of an area where the relief has almost completely lost its original character due to the formation of vast heaps and settlers is the area surrounding the Lazy mine in the Karvina part of the basin (Mulkova 2003).

The spoil tips accumulate material of varied grain size, mostly sandstone, clay, and coal (10–15 %). The production of waste rock in the basin has been regularly monitored only since 1963; therefore, for the older periods, it may only be estimated. It is assumed that since the beginning of mining until 1900, the extraction of rock mined amounted to 20 % of the extracted coal, which was about 12 million tonnes. In later years, until the mid-twentieth century, the rate of waste per tonne of coal was still advantageous and stood at 20–25 %, which resulted from a relatively clean method of operation. Since the 1960s, this ratio was generally higher than 50 % (Havrlant 1979).

Fig. 8.9 Settling tank in the Darkov mine, Ostrava-Karvina Coal Basin, the Czech Republic (Dulias 2012)

In the 1970s, nearly 1 million m^3 of waste rock was mined yearly; some of it was stored on 75 major spoil tips, covering an area of 5.5 km^2 (Mareš 1975). The heaps had different shapes—conical, table, terrace, or irregular—and some of them reached significant heights, even up to 91 m. Most of the forms burned. The biggest dumping ground of mining waste, with a volume of 15.3 million m^3, was in Hermanovice within the limits of the Rudý říjen mine. Other large forms included the central spoil tip Zárubek (11.1 million m^3), the heap in Petřvald in the area of the Julius Fučík mine (7.5 million m^3), and a complex of forms belonging to the John Šverma mine (6 million m^3) (Havrlant 1979). The volume of all heaps at the end of the 1970s was estimated at 120 million m^3, which was about 40 million m^3 more than a decade earlier. In the surroundings of some large spoil tips, deformation of the surface layer of the ground was observed, caused by the pressure of the large mass of waste material on the substrate.

In the period of 1900–2000, the production of waste rock reached 638 million tonnes, including 92 million tonnes until 1946; in the years 1947–2000, it was as many as 545 million tonnes (Martinec and Schejbalova 2004). Of the total production of waste during the mining period (0.65 billion tonnes), about 35 % (approximately 226 million tonnes) was moved to spoil tips (Martinec et al. 2005), with a volume of 141.1 million m^3 (Dopita 1997). Currently, there are 279 spoil tips in the basin. The biggest volume heaps are located in the Ostrava part of the basin (73.9 million m^3, 115 forms), due to the fact that 1 tonne of extracted coal produces 1.07–1.78 tonnes of waste rock. In the Karvina part, this ratio is more favourable

(0.34–0.67 tonne) because the volume of waste rock accumulated on the 143 spoil tips amounts to 45.3 million m^3. In the southern part of the basin, there are 21 spoil tips, which hold 21.9 million m^3 of waste rock. The "burdening" of the surface with mining waste, however, is highest in the Karvina part; up to 4.8 million tonnes are accrued per 1 km^2, which is 2.7 times more than in the Ostrava region and 8 times more than in the southern part of the basin (Schejbalova 2003).

In the anthropogenic relief of the basin, settling tanks (kalište) play an important role. Many of them were created on former fishponds, such as the Kuboň in the area of the Paskov mine. Ponds that occurred in the valleys of the Oder, the Ostravica, the Stružka, and the Olza, and in the pre-mining period constituted distinctive elements of the landscape, and have mostly disappeared or have other functions (Martinec et al. 2005). As for 1991, there were more than 70 settlers in the basin in various stages of filling, with a total capacity of more than 31 million m^3 according to Martinc and Schejbalova (2004) or 28.1 million m^3 according to Dopita (1997). Currently, settling tanks occupy an area of 4.7 km^2, including active ones of 2.8 km^2 with a number of 35, mostly in the Karvina part (26) (Walek and Kostruch 1996; Latová 2003). The production of silt has steadily been decreasing: in 1990, it was 1.5 million tonnes, whereas in 1995, it was 0.2 million tonnes. Post-carbon silts are used in the economy in the energy industry, coking plants, or greenhouses.

Mining activities have disturbed the flow of rivers through the local reversal of their decline. The length of courses covered by subsidence is 115 km, including the Ostravica riverbed (23 km), the Olza (21.5 km), the Oder (15.1 km), and other courses from a few to 10 km (Stružka, Olešná, Karvinský potok, Sušanka, Černý příkop, and Lučina) (Maniček 2003). The increased gradients are mitigated by the construction of drops, such as in the Ostravica riverbed. Very significant changes have occurred in the upper and middle segments of the Karvinski Potok.

In the past 100 years, almost 2 billion m^3 of water was pumped from the mines, of which over 80 % came from the Ostrava region. Closed workings in this part of the basin are still drained in order to protect the Karvina mines against flooding. In the decades after the cessation of mining and dewatering, the groundwater level may rise to an elevation of 210 m above sea level, which may affect geotechnical properties of the bedrock (Atlas map vlivu útlumu hlubinné těžby černého uhlí 2003).

An approximate measure of anthropogenic relief transformation by coal mining may be the amount of extracted minerals (coal and waste rock) per 1 km^2. For the Karvina part, it amounts to 15.7 million tonnes/km; for the Ostrava part, it is 6.6 million tonnes/km; and in the south, it is only 1.7 million tonnes/km (Schejbalova 2003).

Based on the presented data, the rate of anthropogenic denudation in the Ostrava-Karvina Basin was estimated. For the period of intensive operation in the years 1961–1989, the obtained value was 48 mm/year; for 1990–1999, with clearly limited coal production, the rate of denudation was almost the same at 47 mm/year, which well reflects the fact of the concentrated mining impact in an area of a smaller surface. Using data contained in Atlas map vlivu útlumu hlubinné těžby černého uhlí (2003), an average rate of denudation in the area of particular mines in the period 1901–2000 was calculated. It was assumed, following Martinec et al. (2005), that the exploitation ratio is 0.7; the subsidence volume was calculated per impact area and the years of mine operation. The obtained values were in the range of 28–136 mm/year. The highest rate of denudation, above 100 mm/year, was obtained for both old mines in the Karvina part (Lazy: 135 mm/year, Darkov: 103 mm/year) and new mines (9. květen: 136 mm/year, ČSM: 109 mm/year). Overall, in the whole basin, based on accepted calculation criteria, the average denudation indicator value in the last century was calculated at 75 mm/year.

Some areas subsided at a very rapid pace. In the area of the current Darkov Sea in the district of Karvina, mining works were initiated in 1983. By 2005, the surface lowered by 16.5 m—on average, more than 700 mm/year. The topographic maps of 1988 show a built-up area, while the aerial maps of 2000 reveal a vast water reservoir surrounded by a mining wasteland (Santorius et al. 2007). A similar rate of denudation refers to the surroundings of Karvina Doly, with the aforementioned Church of St. Peter from Alcantara, at 680 mm/year.

8.3 The Ruhr Coal Basin (the Ruhr District), Germany

The Ruhr Coal Basin (or the Ruhr District) is situated in the western part of Germany, in the federal state of North Rhine-Westphalia. There are no clearly defined boundaries; the most commonly known name of the Ruhr means the area of "Regionalverband Ruhrgebiet," which is an administrative compound of towns and districts. It is a metropolitan area that consists of 11 cities and 4 districts inhabited by 5.7 million people. The basin area therefore extends from the River Rhine in the west to rural areas of Soester Börde in the east, from the Ruhr River in the south, and to the Münsterland in the north. The extent from west to east is 116 km and from north to south is 67 km. The total area of the Ruhr District is 4,435 km^2 (Bronny et al. 2004).

The bedrock is built of folded, Upper Carboniferous coal-bearing deposits, with outcrops along the Ruhr valley in the southern part of the Basin. Towards the north, rock layers gently plunge into the rocks of Cretaceous age of considerable thickness (Henningsen 1976). Tertiary sediments are found only in the western part of the Basin, while thin layers of Quaternary glacial, periglacial, fluvial, and aeolian deposits spread on the surface (Richter 1996).

The Ruhr District is not very diverse in geomorphological terms. In the south, there are Süderbergland hills and to the north of them is Hellwegzone, with a rather plain character. Local morphological edges are formed by chalk outcrops. Further north, there are low flood plains of the Emscher and the Lippe, partly "broken up" by Tertiary sediments (Liedtke 1993). The majority of the basin is located at altitudes of 100–200 m above sea level, but its southern part rises to 250–350 m.

In the Ruhr Basin, there are three zones, referring to the history of economic development. In the south is the Ruhr zone, where coal mining had its beginnings. Further to the north is the Hellweg zone, referring to the old medieval trade route, with the cities of Dortmund, Bochum, and Essen. The most northern is the Emscher–Lippe zone, with the cities of Gelsenkirchen, Herne, and Bottrop.

Coal mining began in the eighteenth century on the outcrops of Upper Carboniferous coal-bearing rocks in the Ruhr valley. Initially, the operation was carried out on a small scale to the water level, with the engagement of local people. The invention of the steam engine allowed for the drainage of deposits and the extraction of deeper deposits—the zone of more intense mining began to move towards the north, even to the marshy lands in the Emscher valley. The growing demand for coal for the steel industry and railway was satisfied by the increasing extraction and soon the Ruhr became one of three major basins in the world. In 1800, only 0.23 million tonnes of coal were extracted (158 "mines") and 2 million tonnes in 1850; by 1900, it was up to 60 million tonnes (170 mines). The largest production of 130 million tonnes took place on the eve of World War II, in 1939, and in the mid-1950s. In later years, coal production decreased steadily—from 115 million tonnes in 1960, through nearly 76 million tonnes in 1975, to 70 million tonnes in 1980 (Petsch 1982). During this period, about 88 % of hard coal production in Germany came from the Ruhr.

Along with the decreasing extraction, the number of mines also decreased. In 1963, there were 119 active mines; 5 years later, as a result of the restructuring of mining, a coal holding company Ruhrkohle AG, consisting of 7 companies (including 52 coal mines), was established. By 1993, there was only one company left, which, in 1997, was comprised of 13 mines (including the Westfalen, Sophia Jacoba, and Auguste Victoria). In the period of 1969–1997, the output decreased from 84.9 to 35 million tonnes, but the speed of operation increased threefold in the 49 active workings (compared to 360 in 1970) (Sroka 1999). This resulted in a significant increase in the threat to the surface; consequently, costs were incurred for the removal of mining damages.

In 2002, in the Ruhr region, there were 11 active mines. Their total production was 15.5 million tonnes: Ewald-Hugo (1.86 million tonnes), Auguste Victoria (1.64), Prosper-Haniel (1.66), Ost (1.62), Walsum (1.41), Lippe (1.38), Lohberg/Osterfeld (1.27), Friedrich Heinrich/Rheinland (1.24), Westfalen (1.18), Niederberg (1.17), and Blumenthal/Haard (1.05) (Mining Technology 2002). Over the next few years, the number of mines halved. In 2008, out of 7 active coal mines in Germany, 5 were located in the Ruhr and all led their operations at very great depths: Walsum, 1,386 m; Prosper Haniel, 1,197 m; Lippe, 1,330 m; August Victoria, 1,280 m; and Ost, 1,460 m (Harnischmacher, personal communication). Operations are carried out using only the roof-collapse method. The currently operating mines in the Ruhr area include Prosper Haniel in Bottrop and Auguste Victoria in Marlborough (Hamm).

In the period of 1800–1990, a total of 9.5 billion tonnes of coal were extracted in the Ruhr (Meyer 1986; Wiggering 1993a). According to statistics compiled by the author, this constituted nearly 70 % of all the coal mined in Germany in that period.

Long-term and intensive mining activity led to significant changes in the environment of the Ruhr area, especially in the case of the relief and water relations. The main anthropogenic landforms associated with mining are subsidence troughs and spoil tips, as well as embankments of rivers in the polders, while sinkholes, faults, or fissures are generally of minor importance in morphology. Discontinuous deformations in the form of sinkholes formed with particular severity in areas of the former shallow and often illegal mining (Bell et al. 2000). The abandoned pits, shafts, and shallow conveying declines from the 19th and early 20th centuries that were insufficiently filled in after the cessation of mining, and also the illegal excavations dug for heating purposes at the end of World War II, subsided. Sinkholes reached up to 20 m in diameter and had a depth of 15 m. In the last 50 years, discontinuous deformations were also frequently observed, both in areas of shallow as well as deep operation, in the presence of a thick overburden of Cenozoic deposits (Kowalski 2005). Schemes of discontinuous deformations (trenches, drops, fissures, cracks) appearance in areas of shallow operation (up to 100 m) were provided by Kratzsch (1997). From his research conducted mainly on the western side of the Rhine, it may be concluded that fissures are formed especially in the areas of operations of several seams to the common border line of exploitation. Grün (1995), based on statistical analysis of nearly 900 inventoried discontinuous deformations in the area of two mines (374 km^2) conducting deep exploitation, stated that the formation of 88 % of them was also associated with the borders of exploitation (after Kowalski 2005). Nowadays, discontinuous deformations pose a threat, especially in urban areas. For example, in 2011 in Essen, a sinkhole with a depth of 8 m was formed, which swallowed several cars; in 2000, a similar form emerged in the centre of Bochum (Harnischmacher 2007).

The process of surface subsidence in areas located above underground workings was launched in the second half of the nineteenth century, with the introduction of longwall mining. Mining had already entered the Emscher catchment area, characterized by a poorly diversified relief. The average gradient of the river flowing in a wide flat valley was 1.1 %; the middle and lower sections were 0.5–0.4 % (Czaja 1999). In this flat area, even a small lowering, such as a subsidence trough, results in the formation of large overflow areas and wetlands (Brüggemeier and Rommelspacher 1992). At the end of the nineteenth century (1880) in the Emscher valley between Herne and Gelsenkirchen, the surface subsided by up to 5 m (Harnischmacher 2010a). With the prospect of the further lowering of the surface, the extensive areas of plains of the Lippe and the Emscher (583 km^2) were drained by numerous pumping stations to protect these areas against flooding. In the following decades, the area subsided by 5–20 m, but reservoirs are rare due to continuous drainage. The well-known bodies of water in subsidence troughs subject to legal protection due to the natural environment include Lake Lanstrop in Dortmund-Lanstrop, Lake Hallerey in Dortmund-Dorstfeld (Drecker et al. 1995), and Lake Bever near Bergkamen in the north-eastern part of the Ruhr (Duckwitz and Hommel 2002). In the southern part of the basin, in the areas of the oldest coal mining on the Ruhr River, the land subsided by several meters. Current mining in the Ruhr assumes a settling indicator of about 0.9, which is 90 % of the thickness of the exploited deposits.

Harnischmacher (2010a) developed a map of subsidence in the Ruhr, according to the same method as used in this study, based on subtracting two digital elevation models designed for two time periods: 1892 and 1997–2001. The areas of greatest subsidence, of several meters up to 20 m, are found on the border of Gelsenkirchen and Herten in Dortmund, Essen, and Bottrop, which is mainly in the areas of the closed mines Ewald, Consolidation, Minister Stein, and Zollverein and the active mine Prosper Haniel (14 m). The author emphasized the apparent dependence of the subsidence volume on the geological structures, generally oriented from the south-west to north-east: Bochumer Mulde, Essener Mulder, Emscher Mulde and Lippe Mulde. These were the areas of mining concentration where flat, deposited coal enabled the automation and intensification of extraction.

Anthropogenic forms directly resulting from mining activities include spoil tips and surface outcrops. Small pits are associated with the pre-1700 mining industry. These are forms of insignificant size, with a diameter of 10 m and a depth not exceeding 2 m, surrounded by mounds and heaps of overburden. Coal extraction was conducted on outcrops of coal deposits, in the hills of the southern part of the Basin. The inflow of groundwater forced the abandonment of mining pits and the creation of new excavations next to them. As a result, chains of small depressions were formed that are difficult to distinguish from bomb craters from World War II (Harnischmacher 2007).

Spoil tips stand out in the anthropogenic relief of the Ruhr Basin. As of 1980, there were at least 235 of them and they occupied an area of over 25 km^2 (Petsch 1982). In the early years of mining on the outcrops of deposits, waste rock was stored next to mines on the hills in the southern part of the Ruhr District. They were small, conical forms, which are now well-integrated into the landscape and sometimes difficult to identify in forested areas.

From the mid-nineteenth century, when coal mines "migrated" north to the plains between the rivers Ruhr and the Lippe, and with increasing coal extraction, the mining of waste rock increased as well, making spoil tips a distinctive feature of the landscape. In retrospect, in the Ruhr Basin, there are three generations of spoil tips related to the successive stages of mining development—the conical, the table, and the landscape constructions (Benfer and Förster 1990). The oldest spoil tips in the lowland part of the Ruhr were formed in the shape of cones with steep slopes, resulting from the angle of natural rest of waste rocks. Due to the high content of coal (even above 20 %), many heaps burned, and the strong water and aeolian erosion of the steep slopes hindered the introduction of vegetation. The second generation of forms are table or irregular heaps created later, with heights usually not exceeding 40 m; however some, such as in Bottrop, are more than 90 m high. To prevent landslides and earthflows, the slopes were terraced and their inclination averaged 26° (Wiggering 1993b). The amount of waste produced by coal mining in the years 1960–1980 ranged from 50 to 66 million tonnes per year; however, in relation to the annual extraction of coal, the percentage participation of waste steadily increased. In 1960, there was 0.57 tonnes of mining waste per 1 tonne of coal; in 1975, there was 0.76 tonnes; and in 1980, there was 0.9 tonnes (Petsch 1982). The possibilities of waste disposal in the area of indigenous mines were already limited, so central spoil tips were created where waste rock was collected from several mines. These are high-volume forms with medium heights, but they occupy large areas. Since the mid-1970s, third-generation heaps, known as landscape constructions, started to appear. These spoil tips are shaped like natural hills, with flowing slopes, curved lines at the base, and smooth edges and domes (Harnischmacher 2007).

The volume of waste rock mined in the Ruhr was estimated at 2–2.5 billion m^3 (Bell et al. 2000). The majority of mining waste (over 70 %) was collected on spoil tips; the remainder was used as a substitute of construction material or for backfilling in mines. Spoil tips occupy only 0.6 % of the area of the Ruhr Basin, but are located in relatively short distances from one another and constitute important elements of the contemporary relief.

The effects of mining activities in the form of surface subsidence forced the need to engineer the morphology of riverbed channels (Fig. 8.10). Already in the late nineteenth century, the hydraulic declines of the Emscher and its tributaries were reversed, which, in combination with the discharge of large amounts of municipal and industrial wastewater into rivers, caused serious sanitary problems, especially during floods. In 1899, a new sewage system was designed. Riverbeds were concreted and hundreds of kilometres of sewers (approximately 600 km) were brought to them. To maintain sewage and accelerate the drainage of large areas of plains, the bottoms of riverbeds and channels were deepened and straightened, and the areas of large subsidence were embanked (Bell et al. 2000). Currently, work is under way to restore the

Fig. 8.10 Regulated channel of the Emscher River near Dortmund in the Ruhr District, Germany (Dulias 2008)

natural morphology of riverbeds, consisting of the removal of concrete bottoms, filling them in with river sediments, and causing winding courses of rivers and the introduction of vegetation along the banks. A successful example of fluvial morphology restoration may be the Deininghauser Bach, meandering through a broad flood plain near Castrop-Rauxel. A complete restoration of the Emscher catchment is planned for 2020 and will cost about 4.4 billion Euro (Harnischmacher 2007).

The volume of anthropogenic denudation in the area of the Ruhr was estimated by two methods: one based on the volume of extraction and the other using morphometric analysis. Bell et al. (2000) calculated that the extraction of 9.54 billion tonnes of coal (until 1990) led to the formation of voids in the bedrock with a volume of 7 km^3. Taking into account waste rock (2–2.5 km^3), these voids had a volume of at least 9 km^3. In the Ruhr, stowing sands were only used in special cases; this is why it is assumed that the scale of surface subsidence is 50–90 % of the thickness of the exploited deposits. Taking the above into account, it was found that the volume of subsidence may range from 4 to 7.2 km^3. Harnischmacher (2007) calculated on this basis that the Ruhr Basin area subsided during the period of mining at a rate of 4.7–8.5 mm/year. This indicator is several hundred times higher (362–654) than the one calculated by Schmidt (1984) for the Ruhr catchment area, on the basis of the transported suspension at the measuring station at its mouth, amounting to 0.013 mm/year.

Harnischmacher (2010b) calculated the volume of subsidence in the Ruhr Basin by subtracting the digital elevation models for the end of the nineteenth century (1892) and the end of the twentieth century (1997–2001). He established that surface subsidence occurred in both the natural depressions and on the heights of the area. The average decrease of the surface, calculated for individual sheets of digital maps, ranged from 0.51 m for the area within the boundaries of the *Kamen* sheet, located on the eastern border of the Ruhr Basin, to 5.16 m for the area covered by the *Gelsenkirchen* sheet (128.5 km^2), in the central part of the Emscher floodplain. For one of the research areas in the central part of the Basin, occupying about 131 km^2 (no name of map sheet), the volume of subsidence was calculated at 0.695 billion m^3, which gives an average subsidence of 5.27 m; taking into account the period of exploitation, the surface subsidence indicator was achieved with an average value of 33.4 mm/year (Harnischmacher and Zepp 2009). An average subsidence across the area covered by the analysis of changes in altitude (about 2,700 km^2) is 1.6 m. On this basis, it can be inferred that the area lowered at an average rate of 15 mm/year (calculations by the author). Given the accuracy of archival maps, data analysis takes into account differences in height greater than 2 m and lower than 2 m (Harnischmacher and Zepp 2008, 2014).

The following calculations of the intensity of anthropogenic denudation were carried out on the basis of oral information obtained from S. Harnischmacher in 2008 and 2009. In the area covered by the *Dortmund* sheet (133.3 km^2), there were 3 active mines: Minister Stein (1856–1987), Germania (1855–1971), and Hansa (until 1980). As a result of their mining activities, the area subsided by 18.4, 19.6, and 15.6 m, respectively, whereas the deepest subsidence troughs in the area of the Minister Stein mine (25 m) were filled in with waste material and are not reflected in the contour lines of contemporary topographic maps. The volume of subsidence amounts to 0.38 km^3 and refers to an area of 99.5 km^2, which gives an average lowering of 3.8 m—that is, at a rate of 40 mm/year. The volume of anthropogenic landforms (not only on spoil tips) accumulated in the area of 28.8 km^2 is 0.08 km^3; for this reason, the area elevated by an average of 2.8 m, which gives a rate of 29 mm/year. The denudation balance calculated for the whole characterized area (133.3 km^2) is negative and amounts to an average of 23 mm/year. The rate of denudation varied over time because it generally referred to the intensity of the extraction of minerals and is, to some extent, reflected in the following data. In the area of Dortmund, mining activities were completed in 1987. The largest subsidence occurred to the north of the city centre—18.4 m (Harnischmacher 2010a). Twelve years earlier, the maximum subsidence for the same place was 14 m (Sauer 1976). Therefore, in the period from 1976 to 1987, the area subsided by 4.4 m—that is, at a rate of 367 mm/year (author's calculations).

The relief of the three analysed areas of coal mining in Europe—the Upper Silesian Coal Basin in Poland, the Ostrava-Karvina Basin in the Czech Republic, and the Ruhr Basin in Germany—bears clear signs of anthropopressure. Generalizing, until the end of the twentieth century, in the Upper Silesian Basin and the Ruhr, similar amounts of coal were extracted (9–10 billion tonnes); in the Ostrava-Karvina Basin, it was about 6 times less. Due to very significant differences in the surface of mining impact areas (the USCB is at least twice less than in the Ruhr Basin and the Ostrava-Karvina Basin is approximately 5 times less than in the USCB) the rate of anthropogenic denudation is varied. Based on available data, it was calculated that the highest intensity of anthropogenic denudation occurred in the Czech Basin in the last century; in particular mining regions, it amounted to 28–136 mm/year. In the last 40 years, throughout the Ostrava-Karvina Basin, it averaged 47 mm/year. The anthropogenic denudation rate for the area of the Upper Silesian Basin, in the period 1883–1993, averaged 27 mm/year; for individual geomorphological regions, it was in the range of 2–43 mm/year (for the mines, the anthropogenic denudation rate, calculated on the basis of the extraction of coal and waste rock, ranged from 4 to 156 mm/year, at an the average of 21 mm/year). For the biggest mining area in terms of size, the Ruhr, an average value of the anthropogenic denudation stands at 15 mm/year; however, this rate is spatially varied from a few to more than 40 mm/year. From the above, it may be concluded that in these coal basins, the rate of anthropogenic denudation averaged several to several tens of mm/year, which makes it many times greater than the rate of natural denudation. Therefore, in the last century, the relief of the analysed areas was under the outstanding and, in many areas, dominant influence of mining activities.

Coal is mined in nearly 70 countries and world production shows a steady upward trend. In 2009, it amounted to 5.6 billion tonnes; in 2013, it was 7.2 billion tonnes. The pressure of mining on the environment is not going to cease and further transformations of the relief are inevitable. In Europe, coal is mined mostly underground; in the United States and Australia, over 60 % of extraction comes from open-pit mines; in Colombia and Venezuela, almost the whole production is the result of opencast mining. Coal basins, which carry out intensive extraction, are moreover located in different climatic zones: about half of the mining areas are located in different varieties of temperate climate, approximately 20 % in tropical and subtropical zones, and a few percent in equatorial and circumpolar zones. Anthropogenic relief is and will be modelled in varied environmental conditions.

The measure of mining anthropopressure may be, to some extent, the total of coal mining in different countries. Clearly, it is most important in the United States and China, which will continue to pursue intensive coal mining in the short term. Next, there is the United Kingdom and Germany, in which the main period of relief transformation due to mining has already been completed. Coal mining in Poland ranks the country at 6th place in the world. Given that mining activities in Poland have been conducted in the smallest area in comparison to the five preceding countries, it can be concluded that mining anthropopressure in Poland is the largest.

References

Aldorf J, Grmela A, Hrubešová E (2000) The largest breakdown of the shaft in the history of Czech Mining Industry. Zesz Nauk Politech Śl Górnictwo 246:67–75

Aleshina IN, Snytko VA, Szczypek S (2008) Mining induced ground subsidences as the relief-forming factor on the territory of the (Southern Poland). Geogr Nat Resour 29:288–291

Atlas map vlivu útlumu hlubinné těžby černého uhlí v České část o hornoslezské pánve na povrch a životní prostředí. Documenta Geonica, Ústav Geoniky, Ostrava, 2003

Badura J, Wojtkowiak A (1983) Współczesne pionowe ruchy tektoniczne na Dolnym Śląsku w świetle interpretacji danych hydrogeologicznych. Współczesne i geotektoniczne ruchy skorupy ziemskiej w Polsce 4. Ossolineum, Wrocław, pp 239–250

Bell FG, Stacey TR, Genske DD (2000) Mining subsidence and its effect on the environment some differing examples. Environ Geol 40 (1–2) December:135–152

Benfer S, Förster H (1990) Problematyka hałd górniczych w Zagłębiu Ruhry. Zesz Nauk Uniw Jagiell, Prace Geogr 78:79–91

Bilans zasobów kopalin i wód podziemnych w Polsce (2001–2015). PIG Warszawa

Bogda A, Chodak T (1995) Niektóre właściwości fizyczne i skład mineralogiczny osadów poflotacyjnych ze zbiornika „Gilów". Zesz Probl Post Nauk Rol 418(1):415–420

Bronny HM, Jansen N, Wetterau B (2004) The Ruhr Area. Kommunalverband Ruhrgebiet, Essen

Brüggemeier FJ, Rommelspacher T (1992) Blauer Himmel über der Ruhr. Geschichte der Umwelt im Rhurgebiet 1840–1990. Klartext Verlag, Essen

Cabała J (2005) Kwaśny drenaż odpadów poflotacyjnych rud Zn-Pb; zmiany składu mineralnego w strefach ryzosferowych rozwiniętych na składowiskach. Zesz Nauk Politech Śląskiej 1690. Górnictwo 267:63–70

Czaja S (1999) Zmiany stosunków wodnych w warunkach antropopresji. Wyd Uniw Śląskiego, Katowice

Demek J (ed) (1987) Hory a nížiny. Zeměpisný lexicon České Socialistické Republiky, Akademia, Praha

Dopita M (ed) (1997) Geologie české části hornoslezské pánve. Minist. životního prostředí České republiky, Praha

Douglas I, Lawson N (2001) Materials flows for mining and quarrying. In: Munn T (ed) Encyclopedia of Global Environmental Change 3. Causes and consequences of global environmental change, pp 454–461

Drecker P, Genske DD, Heinrich K, Nol HP (1995) Subsidence and wetland development in the Ruhr district of Germany. Land Subsidence, IAHS Publ 234:413–421

Duckwitz G, Hommel M (eds) (2002) Vor Ort im Ruhrgebiet - ein geographischer Excursionsführer 3. Auflage, Essen
Dulias R (2013) Denudacja antropogeniczna na obszarach górniczych na przykładzie Górnośląskiego Zagłębia Węglowego. Wyd Uniw Śląskiego, Katowice
Galas K (2004) Gospodarka surowcami ilastymi barwnie wypalającymi się do produkcji ceramiki budowlanej, kruszyw lekkich, cementu i innych zastosowań. In: Ney R (ed) Surowce mineralne Polski. Surowce skalne. Surowce ilaste, Wyd Inst GSMiE PAN Kraków, pp 383–400
Garlicki A (1999) Złoża soli w Polsce i perspektywy ich wykorzystania. In: Jankowski AT (ed) Perspektywy geologii złożowej i ekonomicznej w Polsce. Wyd UŚ, Katowice, pp 66–75
Gąsowski Z (1994) L`enfoncement du lit de la Loire. Revue de Géographie de Lyon 69–1:41–45
Goudie AS (1993) Human influence in geomorphology. Geomorphology 7(1–3):37–59
Goudie AS (2006) The human impact on the natural environment, 6th edn. Blackwell, Oxford
Goździk J, Dylik W, Szataniak J (2009) O procesach eolicznych w wyrobisku Kopalni Węgla Brunatnego „Bełchatów". In: Dulias R, Pełka-Gościniak J, Rahmonov O (eds) Ekosystemy piaszczyste i człowiek. WNoZ UŚ, Sosnowiec, pp 84–95
Grün E (1995) Analyse und Prognose von Unsteitigkeiten als Folge bergbaubedingter Bodenbewegungen im linksniederrheinischen Steinkohlengebiet. Dissertation. RWTH, Aachen
Harnischmacher S (2007) Anthropogenic impacts in the Ruhr District (Germany)—a contribution to anthropogeomorphology in a former mining region. Geogr Fis Dinam Quat 30:185–192
Harnischmacher S (2010a) Bergsenkungen im Ruhrgebiet. In: Heineberg H, Wieneke M, Wittkampf P (eds) Westfalen regional, vol 2. Wissenswertes und Medien über die Region Westfalen-Lippe. Münster, Aktuelle Themen, pp 124–125
Harnischmacher S (2010b) Quantification of mining subsidence in the Ruhr District (Germany). Geomorph relief proc environ 3:261–274
Harnischmacher S, Zepp H (2008) Mining subsidence in the Ruhr District. Congress I.A.G./A.I.G. Working Groups HILS & IAGeomhaz, Bochum, p 5
Harnischmacher S, Zepp H (2009) Quantification of mining subsidence in the Ruhr District (Germany)—a contribution to Anthropogeomorphology. In: 7th International conference geomorph—ancient landscapes—modran perspectives, Melbourne. Abstracts
Harnischmacher S, Zepp H (2014) Mining and its impact on the earth surface in the Ruhr District (Germany). Z Geomorph, Suppl Issues 58(3):3–22
Havrlant M (1979) Antropogenni formy reliéfu a životní prostředi v Ostravské průmyslové oblasti. Spisy Pedag Fakulty v Ostravě 41, PF v Ostravě
Henningsen D (1976) Einführung in die Geologie der Bundesrepublik Deutschland. Ferdinand Enke Verlag Stuttgart
LeB Hooke R (1994) On the efficacy of humans as geomorphic agents. GSA Today 4(9):217–225
LeB Hooke R (2000) On the history of humans as geomorphic agents. Geology 28:843–846
Hortvik K (2003) Vysvětlivky a komentář k mapám vlivu poddolováni s poklesy dobývání za léta 1961-1989, 1961-1999 a 1990-1999. Documenta Geonica, Ústav Geoniky, Ostrava, pp 47–62
Kasztelewicz Z, Zajączkowski M (2010) Wpływ działalności górnictwa węgla brunatnego na otoczenie. Polit Energ 13(2):227–243
Kijewski P, Downorowicz S (1987) Odpady poflotacyjne rudy miedzi jako potencjalna rezerwa surowcowa. Fizykochem Probl Mineral 19:205–211
Kirchner K, Hradek M (2004) Typy reliéfu Ostravska. Documenta Geonica, Ústav Geoniky, Ostrava, pp 29–37
Klimek K (1996) Aluwia Rudy jako wskaźnik 1000-letniej degradacji Płaskowyżu Rybnickiego. In: Kostrzewski A (ed) Geneza, litologia i stratygrafia utworów czwartorzędowych II. Wyd UAM, pp 155–166
Kowalczyk Z (1964) Analiza wyników badań geodezyjnych nad współczesnymi naturalnymi ruchami powierzchni południowej części Górnego Śląska. PAN Geodezja 1, Kraków
Kowalik S, Gajdowska M, Herczakowska J (2009) Problem ochrony środowiska w górnictwie otworowym na przykładzie Kopalni i Zakładów Chemicznych Siarki „Siarkopol" SA—Kopalnia „Osiek". Bud Górn Tunelowe 2:23–27

Kowalski A (2005) Rozpoznanie i możliwości prognozowania liniowych deformacji nieciągłych powierzchni. In: Kwiatek J (ed) Problemy eksploatacji górniczej pod terenami zagospodarowanymi. Ustroń: 278–291

Kozacki L (1980) Przeobrażenia środowiska geograficznego spowodowane wgłębnym górnictwem węgla brunatnego na obszarze środkowego Poodrza. UAM, Geografia 21. Poznań

Kozacki L (1987) Problemy stabilizacji środowiska przyrodniczego obszarów pogórniczych węgla brunatnego na przykładzie odkrywki "Gosławice" (woj. konińskie). In: Szukalski J (ed) Mat. Gdańsk, III Zjazdu Nauk PTPNoZ, pp 124–136

Kozłowski S (1983) Przyrodnicze uwarunkowania gospodarki przestrzennej Polski. Zakł Narod Ossolińskich, Wyd PAN, Wrocław

Kratzsch H (1997) Mining subsidence engineering (German). 3rd edn. by Deutscher Markscheider-Verein, Bochum

Krawczyk A, Perski Z, (no date) Zastosowanie satelitarnej interferometrii radarowej na terenach eksploatacji rud miedzi w LGOM. home.agh.edu.pl

Krupiński B (1971) Rodzime surowce mineralne w gospodarce narodowej Polski. Wyd Śląsk, Katowice

Kupka R, Szczypek T, Wach J (2005) Morphological effect of 200-years long hard coal exploitation in Katowice. In: Szabó J, Morkūnaitė R (eds) Landscapes—nature and man. Univ Debrecen, Lithuanian Inst. Geol Geogr, Debrecen-Vilnius, pp 95–100

Langer P (2007) Rekultywacja i zagospodarowanie poeksploatacyjne terenów salinarnych. Czas Techn Wyd Politech Krakowskiej 7-A: 309–315

Latová A (2003) Vysvětlivky a komentář k mapě nádrží a odvalů. Documenta Geonica, Ústav Geoniky, Ostrava, pp 91–109

Leszczycki S (1980) Geograficzne studium ekonomiczno-planistyczne. Nad mapą Polski. Prognozy, Perspektywy. Wyd Książka Wiedza, Warszawa

Liedtke H (1993) Die Entwicklung der Oberflächenformen im Ruhrgebeit. Berichte zur deutschen Landeskunde 67:255–265

Łuszczkiewicz A (2000) Koncepcje wykorzystania odpadów flotacyjnych z przeróbki rud miedzi w regionie Legnicko-Głogowskim. Inż Mineral I-VI: 26–35

Maciejewska A (2000) Rekultywacja i ochrona środowiska w górnictwie odkrywkowym. Oficyna Wyd Politech Warsz, Warszawa

Madowicz A (2001) Osiadania terenu na obszarze Jastrzębia Zdroju w latach 1974-1997. Kształt środ geogr ochr przyr obsz uprzem zurb 31:15–21

Maniček J (2003) Mapa ovlivnění vodních toků a vodních ploch dobýváním černého uhlí v české části hornoslezské pánve se zákresem záplav povodně w r. 1997. Documenta Geonica, Ústav Geoniky, Ostrava, pp 63–81

Mareš J (1975) Vliv člověka na životní prostředí Ostrawska. Stud Geogr, ČAV, Geografický ústav Brno

Martinec P, Schejbalova B (2004) History and environmental impact of mining in the Ostrava-Karvina Coal Field (Upper Silesian Coal basin, Czeech Republic). Geologica Belgica 7(3-4):215–223

Martinec P, Schejbalová B, Hortvík K, Maníček J (2005) The effects of coal mining on the landscapes of the Ostrava Region. Moravian Geogr Reports 13:13–26

Maruszczak H (1988) Zmiany środowiska przyrodniczego kraju w czasach historycznych. In: Starkel L (ed) Przemiany środowiska geograficznego Polski. Zakł Narod Ossolińskich, Wyd PAN, Wrocław, pp 109–135

Meyer DE (1986) Massenverlagerung durch Rohstoffgewinnung und ihre umweltgeologischen Folgen. Z. deutsch Geol Ges 137:177–193

Mika W, Paszcza H (2007) Rewitalizacja terenów zdegradowanych w przemyśle siarkowym. Wiad Górn 10:541–550

Mining Technology 2002. www.mining-technology.com/projects/germany

Mizera A (1980) Procesy eoliczne na powierzchni zbiornika „Gilów". Cuprum 2:24–27

Mizera A, Nierzewska M (1996) Metody ograniczania emisji pyłów ze składowiska odpadów „Żelazny Most". Bezp Pracy Ochr Środ Górn 10:42–47

Molenda D (1972) Kopalnie rud ołowiu na terenie złóż śląsko-krakowskich w XVI-XVIII wieku. Inst. Historii Kult Material PAN, Zakł Narod Ossolińskich
Molenda D (1978) Dzieje Olkusza do 1795 roku. In: Kiryk F, Kołodziejczyk R (ed) Dzieje Olkusza i regionu olkuskiego 1, Warszawa
Mulkova M (2003) Detection of changes of the land-use in stoped grounds by means of aerial photography. Intern. Arch Fotogram Remote Sens Spatial Inform Sci 34(7):134–137
Müller J, Ruppert H, Muramatsu Y, Schneider J (1999) Reservoir sediments—a witness of mining and industrial development (Malter Reservoir, eastern Erzgebirge, Germany). Environ Geology 39:1341–1351
Müllerova J, Ides D (2004) Svahové deformace Ostravska. Documenta Geonica, Ústav Geoniky, Ostrava, pp 38–51
Ney R (ed) (2000) Surowce mineralne Polski. Surowce chemiczne. Siarka, Wyd Inst GSMiE PAN, Kraków
Nietrzeba-Marcinonis J (2007) Wpływ rekultywacji leśnej terenów pokopalnianych na wybrane właściwości gleb inicjalnych na przykładzie zwałowiska nadkładu Kopalni Wegla Brunatnego Turów S.A. Uniw Zielonogórski, WILiŚ, Zielona Góra (dissertation)
Nir D (1983) Man, a geomorphological agent. An introduction to anthropic geomorphology. D Reider Publ. Co., Boston and Keter Publ House, Jerusalem, Israel
Nita J (2010) Kamieniołom w krajobrazie i geoturystyce. Prace Kom Krajobrazu Kult 14, PTG, pp 243–251
Nowak A (2008) Ekologiczno-techniczne aspekty procesów pozyskiwania koncentratów cynku i ołowiu. Politech. Krakowska. Kraków (dissertation)
Osmólski T (1971) Historia badań genezy siarki w zapadlisku przedkarpackim w Polsce. Biul IG 246:163–179
Panizza M (1996) Environmental Geomorphology. Elsevier
Pazdur J (1960) Zarys dziejów górnictwa na ziemiach polskich. Wyd Górn Hutn, Katowice
Petsch G (1982) Environmental problems of coal production in the Federal Republic of Germany with particular reference to the Ruhr. Minerals Environ 4:75–80
Pilawska J (1967) Przeobrażanie środowiska geograficznego i rekultywacja w polskich zagłębiach węgla brunatnego. Czas Geogr 38:123–158
Preidl W, Wójcik A (2008) Kopalnia w Posądzy—wpływ płytkiej eksploatacji złóż siarki na deformacje powierzchni. Zesz Nauk Politech Śląskiej, Górnictwo 283:203–216
Pulido-Bosch A, Calaforra JM, Pulido-Leboeuf P, Torres-García S (2004) Impact of quarring in a semidesert karstic area (Sorbas, SE Spain). Environ Geology 46(5):583–590
Pulinowa MZ (1967) Geomorfologiczne metody badania zwałowisk na przykładzie Zagłębia Turoszowskiego. Czas Geogr 38(3):291–297
Radwanek-Bąk B (2001) Trwałość i dynamika przekształceń wywołanych eksploatacją odkrywkową kopalin. Przegl Geol 49:220–224
Repelewska-Pękalowa J (1973) Współczesne procesy morfogenetyczne na zwałach kopalnianych (na przykładzie odkrywkowej kopalni siarki w Piasecznie). Ann UMCS 28(6) B:107–126
Richter D (1996) Ruhrgebiet und Bergisches Land. Zwischen Ruhr und Wupper. Sammulung geologischer Führer 55. Berlin, Sttutgart
Rivas V, Cendrero A, Hurtado M, Cabral M, Gimenez J, Forte L, Delrio L, Cantu M, Becker A (2006) Geomorphic consequences of urban development and mining activities; an analysis of study areas in Spain and Argentina. Geomorphology 73(3–4):185–206
Santorius P, Białecka B, Grabowski J (2007) Środowiskowe i gospodarcze problemy spowodowane degradacją terenów w Górnośląskim Zagłębiu Węglowym. Prace Nauk GIG Górn Środ 1:85–99
Sauer A (1976) Hard coal mining and intensive surface utilization in the Ruhr area. Symp—Environmental Problems Resulting from Coal Industry Activities, Katowice
Schejbalova B (2003) Wysvětlivky a komentář k přehledné mapě dolů—dobývacích prostorů, ložisek mimo dobývací prostory a prognózních území v české části hornoslezské pnie se zákresem produkce uhli, hlušiny a vyčerpané vody. Documenta Geonica, Ústav Geoniky, Ostrava, pp 33–45

Schmidt KH (1984) Der Fluß und sein Einzugsgebiet. Wiesbaden (Wissenschaftliche Paperbacks Geographie)
Skinderowicz B (1974) Wpływ czasu na kształtowanie się dynamicznych niecek osiadania. Prace GIG, Katowice
Skinner BJ (1978) Zasoby Ziemi. PWN, Warszawa
Ślizowski K, Saługa P (1996) Surowce chemiczne. Sól kamienna. In: Ney R (ed) Surowce mineralne Polski Wyd CPPGSMiE PAN, Kraków
Smakowski T, Galas K, Lewicka E (1997) Produkcja i rynek miedzi oraz perspektywy ich rozwoju w kraju. In: Ney R (ed) Surowce mineralne Polski Surowce metaliczne Miedź i srebro. Wyd Inst GSMiE PAN Kraków, pp 139–158
Solarski M, Pradela A (2010) Przemiany wybrany form rzeźby Wyżyny Miechowickiej w latach 1883-1994. Z badań wpł antrop środ 11:78–92
Sprynskyy M, Lebedynets M, Sadurski A (2009) Gypsum karst intensification as a consequence of mining activity (Jaziv field, Western Ukraine). Environ Geology 57(1):173–181
Sroczyński W (1997) Wpływ eksploatacji, przeróbki i przetwórstwa rud cynku i ołowiu na środowisko przyrodnicze. In: Ney R (ed) Surowce mineralne Polski. Surowce metaliczne. Cynk i ołów, Wyd Inst GSMiE PAN, Kraków, pp 155–222
Sroka A (1999) Dynamika eksploatacji górniczej z punktu widzenia szkód górniczych. Stud Rozpr Monogr 58. Wyd. IGSiMiE PAN, Kraków
Starkel L (1988) Przemiany środowiska geograficznego Polski a dzisiejsze geosystemy. In: Starkel L (ed) Przemiany środowiska geograficznego Polski. Zakł Narod Ossolińskich, Wyd PAN, Wrocław, pp 7–24
Stawiarski J, Malinowski N (1983) Możliwości obserwacji wpływów wstrząsów sejsmicznych na terenie KWB „Bełchatów" metodami geodezyjnymi. Współczesne i geotektoniczne ruchy skorupy ziemskiej w Polsce 4. Ossolineum, Wrocław, pp 135–153
Střítežská A, Rafajová A (2004) Kategorizace ploch přírodní složky krajiny a antropogenních tvarů Ostrawska. Documenta Geonica, Ústav Geoniky, Ostrava, pp 69–72
Strzałkowski P (2010) Zarys ochrony terenów górniczych. Wyd Politech Śląskiej, Gliwice
Szczypek T, Trembaczowski J (1987) Wyrobiska po eksploatacji surowców mineralnych w środkowej części Wyżyny Krakowsko-Wieluńskiej. Geogr. Stud Dissert 10:100–112
Szczypek T, Wach J (1991a) Rozwój współczesnej wydmy w warunkach silnej antropopresji. Wyd Uniw Śląskiego, Katowice
Szczypek T, Wach J (1991b) Human impact and intensity of in the Silesian-Cracow Upland (Southern Poland). Z Geomorph NE, Suppl Bd 90, Berlin-Stuttgart, pp 171–177
Szczypek T, Wach J (1993) Antropogenicznie wymuszone procesy i formy eoliczne na Wyżynie Śląskiej. SGP, Poznań
Szewczyk J (2008) Kopalnia Soli „Wieliczka"—80 lat obserwacji deformacji pogórniczych. Gosp Sur Mineral 24(3/2):251–272
Tajduś A, Kasztelewicz Z (2009) Dziesięć atutów branży węgla brunatnego w Polsce czyli węgiel brunatny optymalnym paliwem dla polskiej energetyki w I połowie XXI wieku. Kwart Biul Inform—Wegiel Brunatny 4/69
Taras M, Bernaciak W, Kozek B (2008) Perspektywy podaży oraz prognozowania jakości węgla dla celów energetycznych w planach rozwoju Lubelskiego Węgla „Bogdanka" S.A. Symp Konf 73—Zagadnienia surowców energetycznych i energii w gospodarce krajowej, Ustroń, pp 123–137
Traczyk S (1997) Gospodarka mineralnymi surowcami odpadowymi z górnictwa i energetyki. Przegl Geol 45(5):500–504
Trafas M, Eckes T (2007) Glebotwórcze aspekty oceny utworów sztucznych na przykładzie odpadów po flotacji rud cynku i ołowiu. Geomatics and Environ Eng 1(2):97–110
Traple J, Wilk S (2002) Ekologiczne skutki eksploatacji soli kamiennej metodą otworową w kopalni „Łężkowice". Cz. 2. Historia likwidacji kopalni. Wiertnictwo Nafta Gaz 19/1:237–253
Tyc A (1989) Współczesne procesy krasowe w strefie oddziaływania kopalń olkuskiego okręgu rudnego. Krasi Speleologia 6(XV):23–39

Tyc A (1997) Wpływ antropopresji na procesy krasowe Wyżyny Śląsko-Krakowskiej na przykładzie obszaru Olkusz—Zawiercie. Kras i Speleologia 2. UŚ Katowice

Uberman H, Ostrega A (2004) Sposoby rekultywacji i zagospodarowania zwałowisk nadkładu i składowisk odpadów górniczych. Górn odkr 7–8:80–88

Wachowiak G, Wachowiak A (2004) O kierunku wodnym rekultywacji w polskim górnictwie odkrywkowym węgla brunatnego. Gazeta Obserwatora IMGW 6:17–19

Walek L, Kostruch J (1996) Rekultivace kalových nádrží. MS, OKD, Ostrava

Wanfang Z (1997) The formation of in karst mining areas in China and some methods of prevention. Environ Geol 31(1/2):50–58

Wiggering H (1993a) Bergbaufolgelandschaft Ruhrgebiet. Z deutsch Geol Ges 144:295–307

Wiggering H (ed) (1993b) Steinkohlenbergbau Steinkohle als Grundstoff Energieträger und Umweltfaktor. Geologie und Őkologie im Kontext, Berlin

Wilczyński M (2010) Węgiel brunatny. Gosp Środ 5(20):37–63

Wilkinson BH (2005) Humans as geologic agents. A deep-time perspective. Geology 33:161–164

Wojciechowski T (2007) Osiadanie powierzchni terenu pod wpływem eksploatacji węgla kamiennego na przykładzie rejonu miasta Knurowa. Przegl Geol 55(7):589–594

Wójcik A, Chmura J (2005) Złoża surowców mineralnych i zmiany środowiska naturalnego wywołane przez górnictwo na terenie Bukowna. Górn Inż 29(4):219–236

Wójcik J (1993) Przeobrażenia ukształtowania powierzchni ziemi pod wpływem górnictwa w rejonie Wałbrzycha. Acta Univ Wratisl, 1557. Stud Geogr 59:5–145

Wójcik J (1995) Oddziaływanie form antropogenicznych powstałych pod wpływem górnictwa na środowisko przyrodnicze w Zagłębiu Wałbrzyskim. Przegl Geogr 67(1/2):55–70

Wójcik J (2006) Rozwój rzeźby antropogenicznej pod wpływem górnictwa węglowego w Wałbrzychu i okolicy w latach 1975-1996, w świetle gospodarki odpadami górniczymi. Przegl Geogr 78(1):109–126

Wójcik J (2008) Górnicze zmiany rzeźby terenu rejonu wałbrzyskiego. Landform Analysis 9:339–342

Wójcik J (2011) Przemiany wybranych komponentów środowiska przyrodniczego rejonu wałbrzyskiego w latach 1975–2000 w warunkach antropopresji, ze szczególnym uwzględnieniem wpływu przemysłu. Rozprawy Nauk IGiRR UW 21, Wrocław

Wrona A (1977) Rekultywacja wyrobisk popiaskowych w województwie katowickim. Miasto 11:20–25

Chapter 9
Conclusions

In a thousand-year period of mining in the Upper Silesian Coal Basin, more than 13 billion tonnes of mineral resources and approximately 2–4 billion tonnes of waste rock were extracted from the bedrock. Most of these rock masses (98 %), however, were extracted in the past 100 years, and 94 % of them were associated with the exploitation of coal and stowing sand. The mining impact has covered areas of different geology, relief, water relations, land use and development, and generally of various erosion and denudation potential.

In the USCB, landforms created directly as a result of mining activities occupy an area of almost 150 km^2—half of which are stowing sandpits and one-third are waste heaps. The location of concave forms refers to the lithological characteristics of the bedrock and is associated with certain landforms of higher order—firm rock quarries are present at the thresholds and structural edges, clay-pits are located on high plains made up of cohesive forms, and sandpits—in valleys and depressions with a cover of loose glaciofluvial sediments. Concave forms are present primarily in the area of the North Silesian Upland (50 % of their area, 62 % of their volume) and convex forms (waste heaps) in the South Silesian Upland (48 % of their area, 35 % of their volume) and on the Rybnik Plateau. Most of the direct anthropogenic forms have largely preserved the morphological clarity from the mining period, including many small forms from several hundred years ago (holes and mounds associated with the manual exploitation of ores).

From the geomorphological point of view, the most important effect of mining in the USCB is the creation of indirect anthropogenic landforms, primarily subsidence troughs. Sinkholes that used to be characteristic elements of the relief during shallow mining operations were generally of marginal importance, both morphologically and in terms of the quantity of rock mass displaced due to their formation (about 1 million m^3, or 0.007 % of the total volume). Subsidence troughs cover an area of 1,125 km^2. The theoretical volume of subsidence, calculated on the basis of the extraction of raw materials and waste rock, amounts to 5 billion m^3. The actual volume, calculated from digital elevation models, is 3.3 billion m^3, with the provision that the digital elevation model for the late twentieth century (1993) does not

R. Dulias, *The Impact of Mining on the Landscape*,
Environmental Science and Engineering, DOI 10.1007/978-3-319-29541-1_9

take into account the volume of dozens of square kilometres of subsidence troughs, which had been covered with waste material before 1993.

On the basis of morphometric analysis, it was revealed that the highest volume of subsidence characterizes the Carboniferous zone (40 % of total volume). This area subsided by an average of 3.4 m. The highest average lowering of the surface, however, was noted for the Triassic zone: 4.4 m, including a maximum subsidence of about 35 m. The zone with the deepest subsidence troughs is shaped identically as the course of Saddle Beds. In the Miocene zone, an average lowering of the mining area is 2.1 m. The maximum subsidence here exceeds 20 m, but the zone is dominated by subsidence of less than 2 m (61 % of the mining area).

The emergence of anthropogenic landforms, particularly subsidence troughs, resulted in significant changes in morphometric features of the pre-mining period relief. Vast areas of tens and hundreds of square kilometres have changed their absolute heights; in many areas, new height intervals appeared. In the entire mining area, the surface of areas located above 280 m decreased by more than 42 km^2, with an increase of areas located below 250 m of nearly 34 km^2. For most geomorphological units, the average height of terrain decreased by 0.2–4.5 m; the slightest decrease was noticed on the plains, and the highest was shown in areas with slopes of 1°–5°. In more than three-quarters of the area, there was an increase in altitude, mainly in the range up to 5 m. In general, the relief of plains became more varied; in morphologically varied areas, it was mitigated. In each of the geological zones, the changes of slope inclination are of the same nature: they are expressed by the loss of plains, and an increased share of slopes of all inclination ranges, but to the highest extent, in the range of 1°–3°.

Together with the changes of morphometric features of the relief, the conditions of the circulation of matter have also changed. An increase or a decrease in absolute and relative heights led to changes in the "distance" to the local base levels of erosion, which, combined with changes in slope inclinations, modified the nature and intensity of geomorphological processes. The participants of the circulation of matter are deposits with altered physicochemical characteristics because new lithological deposits appeared on the surface (waste rock, tailings), and natural surface sediments in some areas became desiccated, damp, polluted, devoid of plant cover, and mixed with waste material. Anthropogenic landforms are shaped by contemporary geomorphological processes, such as fluvial or aeolian. On the other hand, their appearance on the relief modifies the course of physicogeographical processes, such as by forcing a change in the course of riverbeds or modifying local wind directions.

One of the most significant changes in the circulation of matter in the USCB is the formation of landlocked basins (mostly in subsidence troughs) and the exclusion of these areas from the fluvial system—more than 122 km^2, or almost 8 % of the mining area, mainly in the Triassic zone, and simultaneously in the zone of the Vistula–Oder watershed. The reversal of the natural terrain gradients led to the direction of runoff to the centre of forms and a clear reduction of its path to the base level of erosion. These are new sedimentary basins.

The interruption in the continuity between slope and fluvial systems resulting from the formation of mine tailings embankments, rail or road embankments, and waste heaps is not without significance for the conditions of the circulation of matter. Disturbances in the circulation of matter in fluvial systems result primarily from changes of the base level of erosion location, the geometry of riverbeds, valley gradients and slope inclinations, the course of watersheds and the resulting changes in catchment areas, and changes in flows and the load of the transported material. In 90 % of the studied catchment areas, slope inclination has increased; in 80 % of them, the energy of the relief has increased. More than 80 % of rivers have reduced their base level of erosion—by an average of 3.4 m in the Vistula basin and 4.8 m in the Oder; three quarters of them have increased their longitudinal declines. The lowering of the base level of erosion mostly affects rivers draining the Miocene zone.

Most rivers and streams in the area of mining subsidence have undergone at least one type of geometry change of their beds (straightening, repaving, embankment, paving of the bottom, reinforcement of the banks, establishment of thresholds, etc.). Some of them (e.g. the Kłodnica, Przemsza, Brynica) have clearly increased flows due to foreign water supply. In many catchments, land use has radically changed and anthropogenic forms appeared. Most of the major rivers have been shortened, while medium and small streams have become longer, especially in their upper segments, and particularly in the Miocene zone of the Oder basin. As a result of mining activities, the erosion potential of the Oder basin rivers has been increased, whereas in the Vistula basin, it weakened. An increased removal of material from the slopes is implemented mainly in small and medium-sized catchments that are poorly developed and not forested. The main rivers are polluted above standards by suspension, mainly coal dust, which is a common occurrence in young alluvial deposits building up the floodplains. Some large and medium-sized rivers transport the matter over short distances, leaving it in flow-through reservoirs.

Lowering of the surface caused by underground exploitation of coal, which is due to the movement of rock mass with a dominant share of the vertical component of motion and without reference to the surface base level of erosion, has been varied in space and time. In the early period of mining, the rate of lowering, calculated on the basis of production volumes, was a few or rarely a dozen mm/year; in the twentieth century, it was tens of mm/year. In recent years, the intensity of denudation in the Silesian Upland has declined, from 28 to 26 mm/year; in the Racibórz-Oświęcim Basin, it has increased from 21 to 30 mm/year, which reflects the direction of movement of mining activity from the uplands to the basins. The concentration of output in a smaller area and a high rate of longwall mining, conducted almost entirely with the roof-collapse method, is reflected in higher values of anthropogenic denudation rates for the last several years. In one third of the mines, they exceed the value of 100 mm/year. The pace of surface lowering due to open-pit mining of stowing sand amounted to several hundred millimetres per year.

Anthropogenic denudation indicators calculated by a morphometric method range from 2 to 43 mm/year, while the rate of anthropogenic aggradation is clearly lower and amounts to approximately 4 mm/year. Denudation balance for geomorphological mesoregions is negative, and the largest value is for the South Silesian Upland (17 mm/year). Individual geomorphological units and catchment areas are highly diverse in terms of denudation intensit; the highest negative balance value was obtained for the Trench from Orzeł Biały at 80 mm/year. Denudation rates calculated by the morphometric method do not differ from the ones calculated on the basis of precise geodetic measurements.

In Poland, no mining area has been within the range of such intense mining activity as the Upper Silesian Coal Basin. In the last half-century, its share in total output of mineral resources in the country amounted to 32 %. At the same time, the total extraction of coal from the mid-eighteenth century to 2009 ranks the USCB as sixth in the world. The effects of mining anthropopressure on the relief of the Upper Silesian Basin are the largest in Poland, Europe, and the world. These changes are regional and long-lasting. The rate of anthropogenic denudation in the area of the Upper Silesian Basin is comparable with the intensity of this process in other coal basins—the Ruhr and the Ostrava-Karvina Basins. Regardless of the method of its calculation, it is many times greater than natural denudation by tens or hundreds of times. To remove such a quantity of material from the bedrock as humans have removed due to mining activities, the natural processes of denudation would need tens of thousands of years.

Index

R. Dulias, *The Impact of Mining on the Landscape*,
Environmental Science and Engineering, DOI 10.1007/978-3-319-29541-1

MIX
Papier aus verantwortungsvollen Quellen
Paper from responsible sources
FSC® C105338
www.fsc.org

If you have any concerns about our products,
you can contact us on
ProductSafety@springernature.com

In case Publisher is established outside the EU,
the EU authorized representative is:
Springer Nature Customer Service Center GmbH
Europaplatz 3, 69115 Heidelberg, Germany

Printed by Libri Plureos GmbH
in Hamburg, Germany